BAUMAN'S FIGHT BOOK

Bauman's Fight Book

Augsburg University Library Cod. I.6.4° 2

Michael Chidester • Jessica Finley • Elias Flatscher
Dierk Hagedorn • Marlene Hurst • Lucien Lee
Charles Lin • Oskar ter Mors • Reinier van Noort
Christian Henry Tobler • Bartłomiej Walczak

Edited by Michael Chidester

HEMA Bookshelf

This book is published as a companion to the facsimile of the Universitätsbibliothek Augsburg's manuscript Cod. I.6.4° 2 produced by HEMA Bookshelf.

A version of CHRISTIAN HENRY TOBLER's translation was previously published in: *Lance, Spear, Sword, and Messer: A German Medieval Martial Arts Miscellany*. Wheaton, IL: Freelance Academy Press, 2022.

Published by HEMA Bookshelf, LLC.
47 High Street #443
Medford, MA 02155
www.hemabookshelf.com

Version 1.0, 2022

ISBN 978-1-953683-26-7 (facsimile)
ISBN 978-1-953683-27-4 (hardcover)
ISBN 978-1-953683-28-1 (softcover)

Typeset in Libertinus Serif Display and Libertinus Sans, which are used under the Open Font License
http://libertine-fonts.org/

Printed by Lightning Source.

Contents

"A book is a snapshot in an ongoing conversation that scholars are having in slow motion."

~ DAN VOGEL

FIGURES AND TABLES

CODICOLOGICAL DESCRIPTION

TRANSCRIPTION AND TRANSLATION

WARP AND WEFT: THE *BAUMAN FIGHT BOOK*'S PLACE IN THE TAPESTRY OF GERMAN *FECHTKUNST*

BAUMAN DAGGER TEACHINGS AND THE AUGSBURG GROUP

Messerfechten From the Augsburg Tradition

Comparing the *Ringen* of *Bauman's Fight Book* to Modern Grappling Styles

Nonlethal Techniques in the *Bauman Fight Book*: Context and Application

Joining the Dots: Additional Comments on the Connection Between *Bauman's Fight Book*, the *Gladiatoria* Group, and Fiore de'i Liberi

Appendix: A Transcription and English Translation of *Vecht-voek*, ms. BPL 3281

This section includes all of the illustrations from V, but doesn't assign them figure numbers.

Acknowledgements

I would like to thank Tracy Seyfert and the artisans at Grimm Book Bindery who produced this beautiful facsimile, and the contributing authors whose work is found in this companion volume.

I also thank the many supporters who took a leap of faith and contributed money to this project with no guarantee that it would produce anything worthwhile.

Bauman Fechtbuch Facsimile Supporters

Scott Aldinger • Nicholas Allen • Michael Allenson • Gabriele Ammann • Grant H. Arnold • David d'Antonio • Cameron Atkinson • Ross Bailey • Jack Baker • Volker Baringer • Paul Barolet • Thomas Belloma • Muli Ben-Yehuda • Simon Berg • Jaron M. Bernstein • Joe Berry • Christopher Bertell • Hildo Biersma • Florian Binder • Devon Boorman • Lorenzo Braschi • Kendra Brown • Daniel Burger • Christian Bürgermeister • Christoph Busche • Charles Buschmann • William Buschur • David Casserly • Chris Castle • Joseph Ceirante • Eris Chen • D. G. Church • Benjamin Conan • Peter Concannon • Christy Conley • Conner Craig • José Antônio Alves da Cunha Coutinho • Myles Cupp • James R. Darling • Dean Davidson • Marla Desat • Jason Drago • Victor Dubernard • Caius Ehmke • Matthias Fassnacht • Jessica Finley • Max Fishman • Elias Flatscher • Philip L. Frank • Jack Fraser • Kevin Frost • Nathan Gede • Rolf Geissbühler • Koen van Gorp • William Grandy • Andrew Greene • Nathan Grepares • Dion Groot • Dierk Hagedorn • Chad Healey • Aaron Himmler • Jacques Hinkley • Nickolys Hinton • James Hodgson • Stéphanie Hoffmann • Steve Hradsky • Marlene N. Hurst • David Huss • J. T. Hutchison • Antti Ijäs • Oliver Janseps • Daniel Jeffery • Peter Kahle • Alexander Kalywas • Rob Kamm • Tea Kew • Don Kindsvatter • Anthony Klon • Hugh T. Knight, Jr. • Arne Koets • Ryan Kohler • Matthys Kool • Nick Kraus • Lucien Lee • Craig C. Ligon • Joseph Lilly • Charles Lin • Magnus Lundborg • Julian Maddox • Tim Magnuson • Łukasz Majewski • Sophie Marshall • Carey Martell • Gilles Martinez • Eric Mauer • Clemens Mayer • R. J. McKeehan • Ken McKenzie • Veronique McMillan • Louis Melancon • Greg Mele • Tracy Mellow • Emanuel Meyer • Matthew Mole • Eliot Mook • Sean Morgan • James P. Morris • Kevin Murakoshi • Keith Nelson • Jacob Norwood • Maxim Nossevitch • Christian Ouellet • Julius Alexander Pedersen • Jennifer Perry • Arjan Peters • Allison Pettengill • Gregory Poulos • Lynda Preston • Brian Price • Brittany Reeves • Nancy D. Reimers • Alexander Ripa • Anthony Rischard • Eetu Röpelinen • John Rothe • Marik Schoenke • Thomas Schratwieser • Till Schultz • A. J. Sedlmair • A. W. Sheppard • Edward W. Sleight, Jr. • Marcin Smolka • Leopold Stoffels •

BENJAMIN STRICKLING • CRAIG TEMPLEMAN • CHRISTIAN HENRY TOBLER • DAN TRUE • JEFF TSAY • CHRISTOPHER VALLI • ASHLEY VOGT • ROLAND WARZECHA • CORNELIUS WEBER • KEN WEBER • SIGMUND AMADEUS WERNDORF • JESSE WHITFIELD • DANIEL WICKSTROM • GUY WINDSOR • CHRISTOPHER WOLFLA • D. E. WYATT • BENJAMIN YEAGER • CONRAD YU • MICHAEL ZIMMERMANN • *Athena School of Arms* • *Noble Science Academy* • *Palm Beach Sword School* • *Saskatoon Bagua* • *SoCal Swords*

PREFACE

There's something magical about holding a Medieval book in your hands. It's a tangible piece of history produced by craftsmen of another age—parchmenters or papermakers, scribes or printers, rubricators, illuminators, tanners, smiths, metalworkers, and binders, each applying their arts to contribute to something that might soon be discarded or might last for a thousand years (sometimes more).

The content of such a book starts with what they wrote and drew in it, which tells us what the person who designed or ordered the book thought was important, but it goes far beyond that—we can learn about how it was intended to be used based on its size and shape and the materials used to create it, and we can learn how it was actually used based on the damage it suffered and the marks left by readers. Notes and doodles on the pages, pages pasted in or torn out, even the crayon scribbles of children—these tell us about the readers, and their relationship to the book.

In this facsimile, I've tried to capture some of this magic and create something that is a bit more than just a fancy book.

High-resolution scans of the manuscript were created by the Augsburg University Library and then carefully cleaned and corrected by HEMA Bookshelf. These were then sent on to Grimm Book Bindery, which printed them on sheets of rag paper similar to the original 15th century paper stock and then collated and sewed them following the current construction of the manuscript (preserving all the errors of previous bindings). Once sewn, the pages were cut and bound in kidskin. Artist ELLIOT RAZO extracted the blind tooling pattern from the preserved remnants of the original 16th century cover, and this design was made into a letterpress stamp to apply to the cover leather.

Clearly this facsimile is a thoroughly modern creation, made by 21st century artisans using 21st century techniques, but it was built based on something old and worn and full of stories.

The purpose of this companion volume is to further illuminate some of those stories. In its pages, you will find articles from some of the leading scholars in the field of historical European martial arts studies that highlight various aspects of the manuscript and the world it came from.

We begin with a a detailed description of the *Bauman fight book* and its history which I authored, and then a full transcription by DIERK HAGEDORN and translation by CHRISTIAN HENRY TOBLER and JESSICA FINLEY. FINLEY and TOBLER also introduce the contents of parts A and B in a paper that discusses the various teachings and explores relationships with the tradition of Johannes Liechtenauer.

Then BARTŁOMIEJ WALCZAK discusses the dagger section and uses it as a lens through which to examine the origins and structure of the Augsburg tradition as a whole. LUCIEN LEE discusses the Messer fencing and proposes connections to other pre-Leckűchner texts. ELIAS FLATSCHER evaluates the wrestling teachings and compares them to modern wrestling traditions. CHARLES LIN and MARLENE HURST present a discussion of the historical context of the fighting teachings of parts A and B and some advice on training them in ways that reflect that context. Finally, DIERK HAGEDORN shows connections between the illustrations in part C and *Gladiatoria*. An appendix contains an edition of another manuscript with similar artwork by REINIER VAN NOORT and OSKAR TER MORS.

It's my hope that this book will serve as a worthy update to the edition of the manuscript that GRZEGORZ ŻABIŃSKI and BARTŁOMIEJ WALCZAK released 20 years ago, *Codex Wallerstein: A Medieval Fighting Book from the Fifteenth Century on the Longsword, Falchion, Dagger, and Wrestling*. Theirs was an important work that introduced many people to this manuscript. May this little book do the same.

MICHAEL CHIDESTER
Medford, MA

A Note on Names

This manuscript is most commonly known in the HEMA community as "Codex Wallerstein", and this probably stems from a misreading of FRIEDRICH DÖRNHÖFFER's 1909 monograph on Albrecht Dürer. While there were ten fencing manuscripts in the Öttingen-Wallerstein Library (kept at Castle Wallerstein until 1841), Dörnhöffer was only interested in the one that was copied by Dürer (i.e. our manuscript); this left him was free to refer to it as "the manuscript from Wallerstein" or just "Codex Wallerstein".

This reference was misconstrued as a proper name when DÖRNHÖFFER's photos of the manuscript illustrations were scanned and placed online in ca. 2000 (on the HACA site and elsewhere). The name was then used by ŻABIŃSKI and WALCZAK in their 2002 book, which cemented it in the minds of most.

Obviously, this name is painfully imprecise once you know its origins—nine of the ten fencing manuscripts previously in the Wallerstein collection are still kept together in the Augsburg University Library, and they all have equal claim to such a designation. In his 2014 German edition, RAINER WELLE proposes the name *Baumanns Fecht- und Ringkampfhandschrift* ("Bauman's fencing and combat-wrestling manuscript").

That's a bit of a mouthful, though, so in this book we're taking our cue from DIERK HAGEDORN and going with *Bauman's fight book* (BF).

1549

Von baumans

108

Fechtbuch.

Codicological Description

Michael Chidester
Wiktenauer & Athena School of Arms

While this manuscript has received considerable attention from scholars and extensive codicological descriptions have previously been published in German,[1] it has been largely neglected in English.[2] I aim to remedy this with the following description and discussion of the physical characteristics, history, and contents of the manuscript.[3]

Augsburg, Universitätsbibliothek • paper, 110 • 1420/1465–70 • Bayern

Cod. I.6.4° 2[4] is a paper manuscript with 110 leaves in quarto format, measuring 140 mm × 210 mm. The page block has three sewing supports. It is bound in a 20th century brown leather cover with fragments of the former 16th century blind-tooled leather cover mounted on it.

The manuscript shows extensive wear and damage. Several leaves, including 1, 2, 3, 87, 88, 107, and 108, show signs of significant repairs. The corners of the leaves are rounded and worn, especially in the first quire. Finally, the manuscript shows evidence of having been trimmed down from its original size, and it may have been bound from Bolognese *recute* or chancery sheets (indicating an original page size of 157 mm × 225 mm).[5]

1 Dating and Origin

BF is a compilation and conflation of at least two different source manuscripts with independent origins.

---

[1] Early work includes Wassmannsdorff 1870 and Dörnhöffer 1907/1909, both of which discuss the manuscript purely as an appendage to their studies of Albrecht Dürer's treatise (AD), whereas Wierschin 1965, Hils 1985, Bodemer 2008, and Leng 2008 offer relatively brief profiles as part of their much larger catalogs of fencing treatises. The first dedicated work on the manuscript was Hils 1991, an analysis written as a companion to a microfilm reproduction, but this effort was dwarfed by the most recent publication on the subject, Welle 2014, a masterful and wide-ranging examination from almost every conceivable direction.

[2] The only full edition in English is Żabiński & Walczak 2002, whose brief discussion of the manuscript in the introduction accepts the assertions of Dörnhöffer and Hils uncritically. Tobler 2022 offers a partial translation but little by way of codicology.

[3] As I have not had the opportunity to examine the manuscript in person, this description seeks to combine my insights and impressions from studying high-resolution digital scans with the prior research of Hils and Welle (who have had that opportunity).

[4] Or *Bauman's Fight Book* (BF).

[5] Welle 2014: 27.

Figure 1: The *mise-en-page* of part A (left) and part B (right) (BF, ff 32v–33r).

Watermark analysis shows that parts A and B are written on Piedmontese paper; part A's paper dates to the 1460s–70s, while Part B's paper is from 1464–65.[6] Furthermore, details of grooming and clothing construction in the artwork suggest a rather tight range of 1465–70.[7] The text is written in a Central Bavarian dialect, possibly in the border area between the Central and Northern Bavarian language areas, and details of the script suggest scribes trained in the second or third quarter of the 15th century.[8]

Of the two sections, it seems likely that part B was created before part A, which was then added to the blank sheets of the last quire of part B (subsequently adding a sixth quire to finish that material).[9] It's possible that they were created in the same workshop in direct sequence, but in either case the different *mise-en-page* (Fig. 1) shows that they represent an extended production unit rather than a single continuous work.

[6] Welle 2014: 29–31; ff 1–2, the end of part B, are on the same paper as part A.

[7] Ibid: 74–76.

[8] Ibid: 69–69.

[9] Evidence for this includes the facts that the illustrations on 1r–2r are in the style of part B, and that the scribe of part A seems to have added a note to the top of 56v.

Part C is written on paper from norther or central Italy with watermarks from 1419–20.[10] It has no text to analyze, but Welle assigns it to the Upper Rhine region based on the art style.[11] Part C is a separate production unit from A and B.

Finally, Paul Hektor Mayr added an additional bifolium after he purchased the manuscript in 1556.

2 History and Provenance

BF's circumstances of creation and early history are unknown. The wear on the corners of the leaves suggests that it was not bound initially and saw use as a booklet.[12]

Around the turn of the 16th century,[13] parts A and B seem to have passed through the hands of Nürnberg artist Albrecht Dürer.[14] He produced a redrawing/reimagining of the wrestling in Part B, which was later included in the compilation manuscript AD. As part of this copying process, Dürer placed red chalk marks and notes next to certain illustrations.[15] He also produced drawings of sword, dagger,

Figure 2: A diagram of the components of BF.[16] B/A are an extended production unit, and C is a separate production unit. Each existed as a separate usage unit before being bound together in the 16th century, after which they have been a single usage unit.

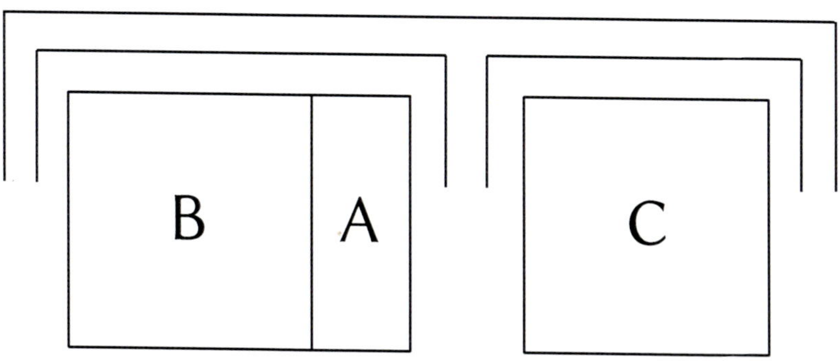

[10] Welle 2014: 32–34.

[11] Ibid: 101.

[12] For a discussion of booklets, see especially Robinson 1980 and Hanna 1986. More recent elaborations of this model can be found in Kwakkel 2002 and Kwakkel 2012.

[13] Hagedorn 2021: 13.

[14] AD doesn't include any material from part C, so it might not have been added to parts A and B yet.

[15] Welle 2014: 89.

[16] Based on standards established in Kwakkel 2002 and Kwakkel 2012.

and Messer plays that may have been inspired by Part A, but the relationship is not as clear as with the wrestling.[17]

The earliest date in BF is 1549, written inside the cover along with "Vonn Baumans" and the number 108. HILS suggests the name refers to the Baumann family of Augsburg, and specifically the mercenary Michael Baumann (fl. 1471–1540),[18] but there is no evidence for this apart from the coincidence of naming. There are 108 sheets in parts A–C, so we may surmise that they were already grouped together at this time. Parts A and B may also have been conflated before 1549.[19]

The first known owner was Paul Hektor Mayr, an Augsburg patrician and avid collector of military artifacts and memorabilia.[20] According to a note on folio 1r, he purchased the manuscript on 26 January 1556. He added a statement to be beginning of each section of the manuscript recording the number of plays, and also added the final two leaves of the current manuscript (ff 109–110), including his index on 109r. He also reproduced the majority of the illustrations in parts A and C in his giant fencing treatise compendia (ignoring the text in this process).

It's likely that Mayr had the manuscript rebound after purchasing it; the sheets inside the covers contain notes predating his ownership (as noted above), but they could have been recycled from the previous binding. Mair added red chalk notations to many pages, and these are out of sequence in the current binding.[21] This is the other point when Parts A and B could have been intermixed.

After his execution on 10 December 1579, Mair's vast collection was inventoried by the city of Augsburg and then liquidated in various auctions starting on 2 May 1580. Unfortunately, no itemized catalog of his library is known to exist, just the item of 26 "fencing and other books". [22] These were purchased by Markus Fugger (1529–1597) for six guilders; since there were other much higher bids, WELLE suggests this might be evidence that Mayr wasn't the city's only corrupt official.[23]

Markus Fugger left some of his library to his grandson Marquart (1595–1655), including "thirteen large and small fencing books... several without words...",[24] and these passed to the library of his brother-in-law Ernst II (1594–1670), Count of Öttingen-Wallerstein, in 1653.[25] Both cover sheets bear purple library stamps from the 'Princely Öttingen-Wallersteinish Library'.[26]

[17] HAGEDORN 2021: 10–13.

[18] HILS 1991: 27.

[19] Given that the older complete foliation, which assigns all pages to their current positions, bears some similarity to the hand that wrote "1549" in the cover.

[20] For information about Mayr's extensive collections, see KIRCHHOFF 2020.

[21] WELLE 2014: 26.

[22] IBID: 22.

[23] WELLE 2014: 22, note 64.

[24] *Dreyzehen gross vnd kleine fecht bücher... mehrer iheil ohne buechstabn;* quoted from WELLE 2014: 22–23.

[25] WELLE 2014: 23.

[26] *F. Öttingen Wallerstein'sche Bibliothek.*

The next mention is in a catalog prepared by archivist J. C. KÖHLER in 1837, where it is the only member of this group of manuscripts that is described. Around this time, the signature "j. h. 10 e 57" was recorded in pencil on the cover sheet.[27]

In ca. 1870, WILHELM FRIEDRICH LÖFFELHOLZ (1809–1891), Baron of Kolberg and chief archivist, cataloged all of the fencing and dueling treatises in the Öttingen-Wallerstein collection. He wrote "Fechtbuch" and "Ringbuch" inside the cover in ink, and this also seems to be when the new signature "Cod. I.6.4° 2" (which it continues to use today) was added in pencil by a different hand.

On 7 May 1935, a collection of manuscripts from the Öttingen-Wallerstein library (including BF) were offered for sale by antiquarian book dealer Karl & Faber, but the only one that sold was GE. Then in 1980, the state of Baravia purchased the

Table 1: The contents of BF.

B	1r	"Arma Gladiatoris"
	1v–2r	Dueling scene *(2v blank)*
A	3r–14v	Unarmored two-handed sword
B	15r–20v	Unarmored wrestling
A	21rv	Unarmored two-handed sword
	22r–28v	Unarmored dagger
	29r–32v	Unarmored Messer
B	33r–71v	Unarmored wrestling
	72r–74v	Dealing with prisoners *(75rv blank)*
C	76r–80v	Unarmored two-handed sword
	81r–95v	Armored dueling *(92rv blank)*
	96r–98v	Unarmored longshield dueling
	98v–100v	Unarmored wrestling
	101r–102v	Unarmored two-handed sword
	103r–108r	Armored dueling
	108v	Marriage proposal scene
N/A	109r	Index by Paul Hektor Mayr *(109v–110v blank)*

[27] WELLE 2014: 21.

entire library and donated it to the University Library of Augsburg, which is the current home of the manuscript.[28]

A final note on the history of the manuscript is that it was rebound in 1989 as part of conservation activities (and a sticker was affixed inside the back cover). The old cover leather and endpapers were preserved and mounted in the new cover. The fragments of a parchment sleeve that previously protected the worn and damaged cover were also preserved separately.

3 Contents

BF contains no title of any sort and its authorship is unknown. It can be divided into three major sections, usually labeled Parts A, B, and C (see Table 1).

Part A includes plays of the sword in two hands, the dagger, and the Messer. Part B is primarily devoted to wrestling, apart from a few plays of dealing with prisoners and four pages containing unrelated illustrations at the end. Due to a very early error in binding, Parts A and B are intermixed in the manuscript (see below).

Part C is a miscellanea including armored dueling with spear, sword, and dagger, as well as unarmored fencing with the two-handed sword, dueling with long-shields, and wrestling.

4 Collation and Foliation

The collation of BF, which is composed of 110 leaves in seven quires without fly-leaves, can be represented according to BRADSHAW's model as follows:[29]

$$\mathrm{I}^{14}\ (\mathrm{f}\ 14) + \mathrm{II}^{24}\ (\mathrm{f}\ 38) + \mathrm{III}^{12}\ (\mathrm{f}\ 50) + \mathrm{IV}^{12}\ (\mathrm{f}\ 62) + \mathrm{V}^{12}\ (\mathrm{f}\ 74) + \mathrm{VI}^{34}\ (\mathrm{f}\ 108) + \mathrm{VII}^{2}\ (\mathrm{f}\ 110)$$

Part A consists of one septernion (quire of seven) and one sexternion (quire of six), part B of four sexternions, and part C of a single massive quire of seventeen folded sheets. The final separate bifolium was added by Paul Hektor Mayr.

Folia 1 and 14 were originally a single bifolium but were subsequently separated, and more recently they were rejoined with a strip of paper. Conversely, the folio pairs 75/108 and 91/92 in quire VI were not originally conjoined: 75 and 92 are a different paper stock than 91 and 108, and were likely added during the 16th century binding to make the singletons 91 and 108 into stable bifolia.

As has been indicated already, it seems certain that part B was produced before part A and they were subsequently reversed in the first binding; the correct order of quires would then be IIa, III, IV, V, I, IIb (Fig. 4). Furthermore, the quires of parts

[28] WELLE 2014: 23.

[29] For a discussion of different methods of symbolizing manuscript structures, see DOROFEEVA 2019. This paper will not attempt the comprehensive collation map she proposes, primarily because the small page format doesn't do it justice.

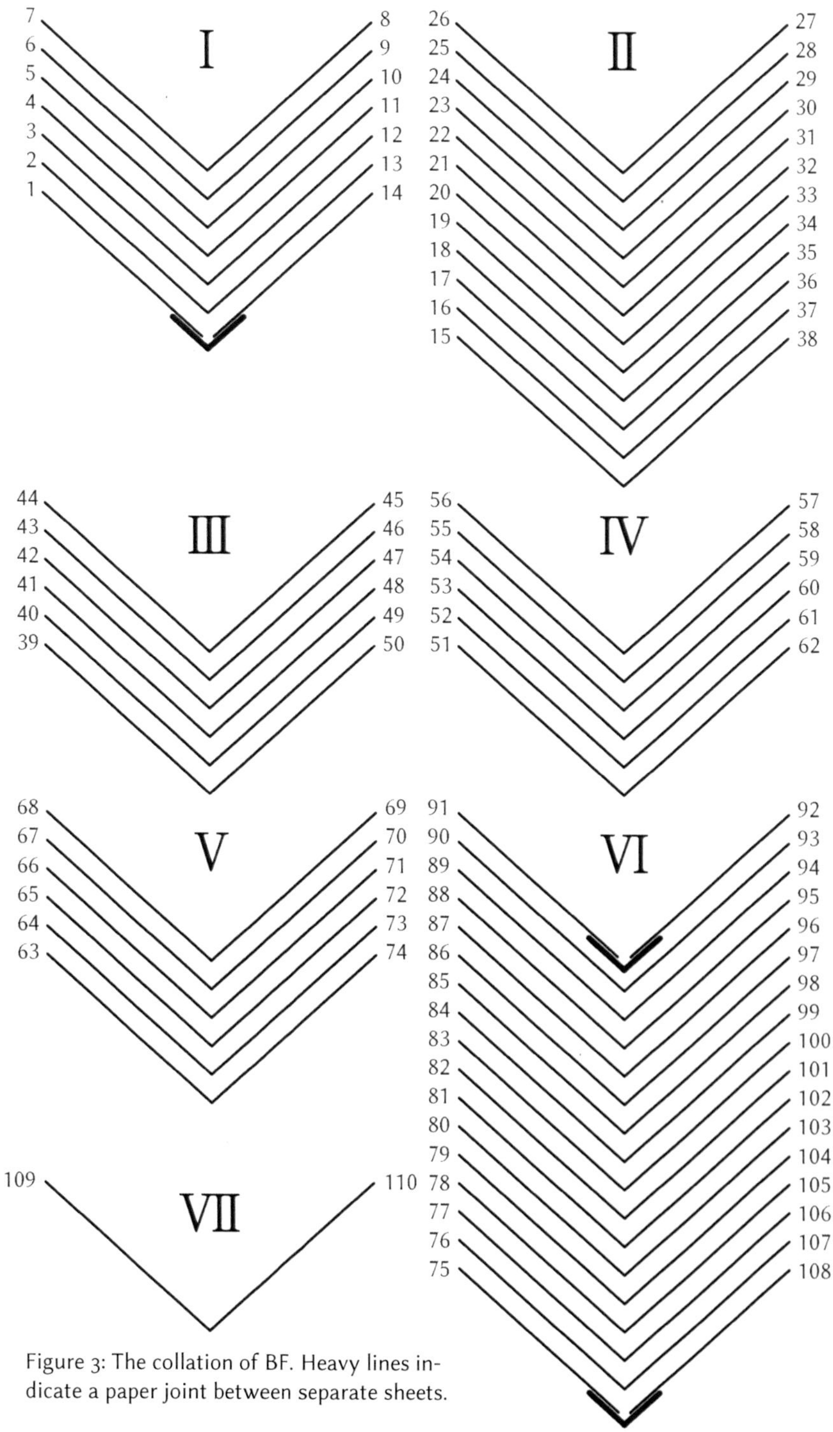

Figure 3: The collation of BF. Heavy lines indicate a paper joint between separate sheets.

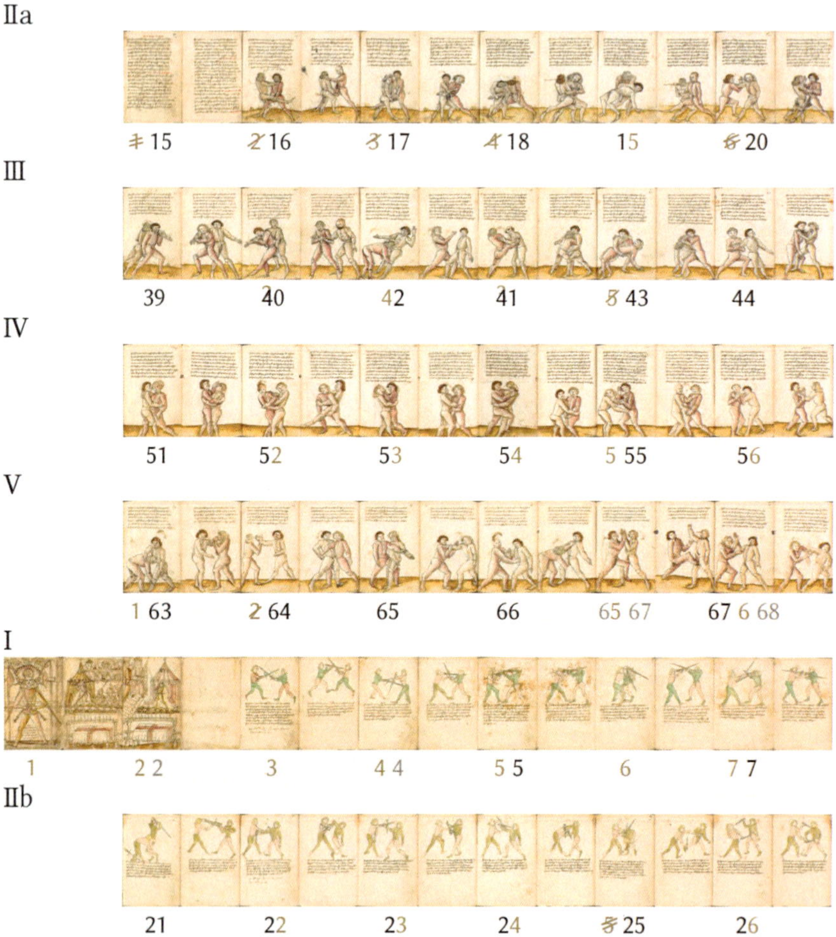

Figure 4: A reconstruction of the original structure of parts A and B, with the fold lining up with the center of each quire. The three paginations are also included in different →

A and B were disarranged in the 16th century during a rebinding (possibly Mayr's). The second quire of part A ended up nested in the first quire of part B, creating a double-length quire and wreaking further havoc on the order of contents in the manuscript. See fig. 4 for a diagram correcting these errors.

A less obvious disordering is the switching of bifolia 42/47 and 41/48. RAINER WELLE demonstrated through analysis of the structure of the wrestling teachings that these pages should be reversed. This confusion of pages happened very early in the history of the manuscript, since even the first foliation (see below) numbers them in their current positions.

IIa

33 34 35 36 37 38

III

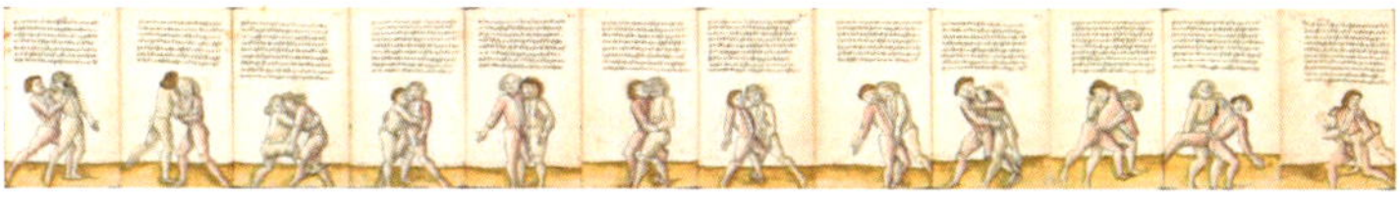

45 46 47 48 49 50

IV

57 58 59 60 61 62

V

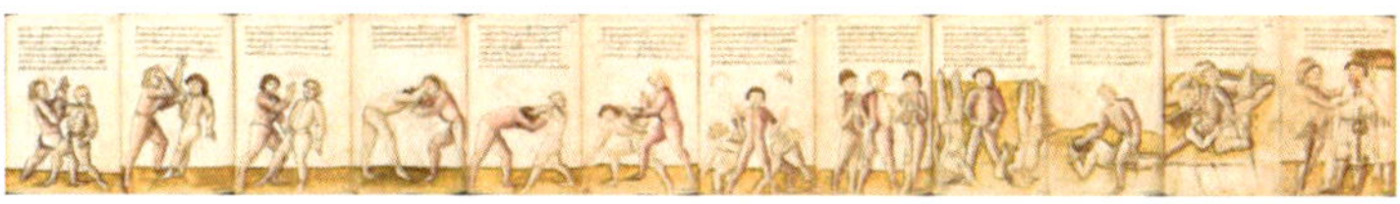

68 69 69 70 70 71 71 72 72 73 73 74

I

8 8 9 10 11 12 13 14

IIb

27 28 29 30 31 32

colors: the brown numbers represent the first foliation, the black numbers represent Mayr's effort, and the grey numbers represent the 19th century corrections.

The leaves in part C (quire VI) also appear to be disordered, but unlike the other material, there's no clear solution to set them right. Both Hils[30] and Welle[31] propose solutions for this problem (involving rotating and rearranging leaves) before acknowledging that it's possible to reduce the issues but not to eliminate them. The disorder was probably already present in the source that was copied for part C.

[30] Hils 1991: 19–21.

[31] Welle 2014: 56–58.

The manuscript shows no obvious evidence of material lacunae, but DÖRNHÖFFER speculates that folia are missing from parts A and B.[32] Part B seems complete,[33] but it is plausible that a leaf was lost from the weapons segment of quire II, which would turn part A into two equal septernions. Since part A doesn't bear a coherent sequence of plays, a leaf might be inserted into it at any point, though losses usually occur at the outside or center of a quire. A loss from the outside of the quire would include two sword and two Messer plays, whereas a loss from the center would include four dagger plays. Both of these scenarios are plausible, since AR (which drew heavily from BF) includes one sword play, two Messer plays, and nine dagger plays which are absent from BF.

BF received folio numbers from at least three different owners (Fig. 4).

The oldest foliation is only found in parts A and B. It is likely from the 15th century and was probably done before they were bound together into a book. Rather than numbering the front of each page, as is typical, only the first half of each quire is numbered (the pages before the center fold). Further, the numbers start over for each quire, beginning at 1 each time. For example, folia 1–7 are numbered but 8–14 (the other half) are not, and then 15–20 are numbered 1–6 again.

The second foliation (and the first continuous one) is in a 16th century hand and was probably made by Paul Hektor Mayr (but possibly the previous owner).[34] He attempted to salvage the older numbers and incorporate them into his scheme, but where he couldn't, he simply crossed them out. For example, the original number on folio 19 was a Medieval 5 glyph, which apparently looked enough like a 9 to Mayr that he merely wrote a 1 next to it. On the other hand, when he came to folio 25, which had a similar 5, he crossed it out and replaced it with a contemporary 5 glyph. Mayr's foliation skips folio 67 and assigns that number to 68, and as such it is off by one from folia 68–109 (and also doesn't number 110).

A third set of folio numbers was added in pencil in the 19th century, probably by the same hand that added the signature "Cod. I.6.4° 2" inside the cover. This foliator corrects Mayr's errors on 67–109, adds a folio number to 110, and adds an additional number to 4 and 8 (probably because those two folia used older numeral forms and were not satisfactory to the writer).

There is a final sequence of indicators worth mentioning, which again seem to come from the pen of Mayr. 6r has the letter 'd', and then 7v–14v and 21rv have a continuous sequence of 'e'–'v'. The dagger is lacking these letters, but in the Messer, 31v and 32v bear the letters 'f' and 'h', which correspond to their positions in the section (6th and 8th).

5 Writing and Decoration

The manuscript is written by two hands in a non-calligraphic practical bastarda

[32] DÖRNHÖFFER 1910: X. He assumes that it was the sole source for AD.

[33] WELL 2014: 47.

[34] IBID: 35.

(corresponding to parts A and B); WELLE characterizes both scribes as skilled and accomplished.[35] Scribe B begins with a dark ink but switches to a lighter ink beginning on 61v, and this lighter ink is also used by Scribe A.

The text of part A is written in 4–8 lines on the bottom half of each page, while the text of part B (where there is text) is written in 4–10 lines. The exception to this is the first folio of part B, which has 22 lines of text on 15r and 23 lines on 15v. There is no original text in part C.

Additional annotations in Paul Hektor Mayr's hand can be found on 1r, 3r, 16r, 22r, 29r, 33r, 80v, 98r, and 100r, and he wrote the index on folio 109r. There are also single-word additions in red chalk, probably by Albrecht Dürer, on 69r, 71v, and 73v.

There are no decorative borders or Lombards in BF, though there is a large bastarda 'I' on 15r. The titles on 15rv are rubricated, and there is other minor rubrication in the form of red slashing and underlining on 15r–16r.

6 ILLUSTRATIONS

There are 209 illustrations in the manuscript by five artists. Artist A drew the 48 figures in part A. Artist B drew the 85 figures in part B. Part C's 83 figures were drawn by at least three artists: C1, C2, and C3.[36]

The artwork in part A began with a rough charcoal outline, followed by pen and ink and then watercolors. (There's no way to determine whether these steps were all performed by the same artist, but we will assume they were.) WELLE sees indications that Artist A was trained in the Gothic art movement.

Artist B illustrated folia 1r–2r in addition to the entire wrestling section. Like Artist A, he began with a charcoal outline, then inked the final figure, and finally applied watercolors. Artist B might have been trained in the Augsburg woodcut style, though it's not a perfect match.[37]

Artist C1 drew the armored dueling figures which make up the majority of part C. He drew confidently with no preliminary sketching, and it's possible that he was working from a pattern book.[38] His watercolors are generally much paler than the other artists.

Artist C2 drew the unarmored fencers in part C, including the sword, wrestling, and dueling with the longshield. He may also have been working from a pattern book, and uses very bright colors.

Artist C3 only worked on the final set of illustrations (f 108v), an apparent courtship scene.

[35] WELLE 2014: 61.

[36] RAINER WELLE offers an extensive art historical analysis in WELLE 2014: 70–105, and most of the discussion in this section is a distillation of his analyses and conclusions.

[37] WELLE 2014: 83. He cites a 1929 monograph by HELLMUTH LEHMANN-HAUPT, but I have not found a copy of it yet.

[38] See DIERK HAGEDORN's article in this volume (pp 195–209).

Transcription and Translation

Dierk Hagedorn (transcriber)
Hammaborg – Historischer Schwertkampf

Christian Henry Tobler (translator)
Selohaar Fechtschule

Jessica Finley (translator)
Ritterkunst Fechtschule

Editor's note: The layout of this section ought to be easy enough follow, but to be clear: when you open this section, the folio number and the transcription by Dierk Hagedorn *appear on the left-hand page and the translation by* Chistian Henry Tobler *or* Jessica Finley *appears on the right-hand side.* Tobler *is the author of the translation of the sword, Messer, and dagger sections, and* Finley *is the author of the translation of the wrestling section. If you're unsure who translated the part you're reading, look to the top of the page.*

1 Transcription Guidelines

The transcription corresponds to the original to the letter including line breaks. Abbreviations are resolved and marked by italics. Long and round forms of the letter 's' are uniformly represented as 's'. Ligatures are not retained, with the exception of 'ß'. Special characters or diacritical markings such as a diaeresis above a vowel (ä, ö, ü, ÿ) have been retained. Sometimes such a diacritical sign could not be recognised unambiguously as a true diaeresis (two individual dots) but rather as a curved line that looks like a reversed tilde or a demicircle.

As was customary at the time, an initial vowel 'u' was frequently written as 'v'; the transcription retains this original 'v', and also the *Umlaut* version 'v̈'. Reciprocally, a 'u' was also used as the consonant 'v', pronounced as 'f' in German.

The capitalisation, although scarce, reflects the original.

Letters, words or passages that were cancelled in the original are represented by ~~crossed out~~ text, and in some instances there are short underlined passgages.

There is hardly any rubricated text, only some headlines and initial letters in the wrestling sections are written in red which was kept for the transcription.

CVR 1549
Vonn Baumans
108

j. h. 10.
e 57[1]

<u>Fechtbuch</u>.

<u>Ringbuch</u> [?][2]

Cod. I.6
4°.2[1]

o
II[1]

1r vber-
khumben Im
1556 Jar am
26 Januarj
paulus hector
mair zugeherig

1v–2v *No text*

3r leng

It*em* so du mit aine*m* vichst vnd zu im kumst an das sw*er*t das ir paid an hapt gepu*n*d*en* so reck dein arm*m* vn*d* dein sw*er*t lanck von dir vn*d* secz dich mit dem leib nider in dy wag vn*d* sich dastu leng vn*d* masß in dem swert habst so magstu arbait*en* vn*d* wer*en* alles das das nottürft ist dy leng das ist dastu hint*er* deine*m* sw*er*t stest vn*d* reckst dich dy masß ist dastu nider stest als hie gm*a*lt stet vn*d* mach dich klain mit dem leib so pistu groß im sw*er*t

Der stennt Im Lanngen schwert
Senndt 24 bar[4]

[1] In pencil.
[2] Mostly effaced.
[3] Library stamp.
[4] From Paul Hektor Mair's hand.

1549
Von Baumann's

<u>Fencing book</u>

<u>Wrestling book</u>

Received in the year 1556 on 26th January.

Belongs to Paul Hektor Mair.

No text

Length

Item, if you fight with another and come to him at the sword so that you have both bound, then extend your arms and sword long before you and set your body low in the balance such that you have length and measure in the sword so you can work and defend as needed. The length is that you stand behind your sword and extend yourself; the measure is that you stand low, as depicted here, and make yourself small with the body, so you are great with the sword.

(Of stances in the bare long sword, there are 24.)

3v masß

Item alstu ainem an pinczt an das swert so slach im lanck ein mit dem ort zu dem gesicht vnd wint im mit der kurczen sneid ein in das angesicht als hie gmalt stet so magstu im ein reissen mit dem knopf oder mit gewappenter hant in das gesicht stossen

4r swech

Item so dw ainem an pinczt an das swert so wart ob er waich oder hert seÿ ist er hert so wint im in das an gesicht als vor geschriben stet ist er aber waich so such dÿ swech seins swercz vnd wint im ÿber auf dein tenke seiten als hie gmalt stet so magstu im nach dem kopf snappen vnd dy plesß suchen

4v sterck

Item haustu ainem oben ein in das swert vnd er helt starck wider so uar auf in dÿ höch mit deinem swert hinder-sich an seinem swert gegen seiner swech vnd das dein gehilcz an seinem swert stee vnd wint im dy kurcz sneid an den hals als hie gmalt stet so magstu in mit gewalt auf dy erden ziechen

5r vor

Item so du ainem in das swert pinczt vnd er slecht dir starck ein zu dem kopf so versecz im mit der zurczen sneid vnd dring in in so musß er slachen slecht er den zu der anderen seiten so lege im dein swert auf sein tencke agssel als hie gmalt stet so slechstu in an ein or das haist ein gelegt vnd vnd haist uor

5v nach

Item so dir ainer oben starck ein haut so wart vnd versecz im dy hew mit der kurczen sneid ist er dann als pehent vnd haut ee wen du im ein magst legen so lasß in freÿ hawen vnd uall im mit der kurczen sneid auf sein swert als hie gmalt stet so pinctzu in vnd slechst in an das or vnd ge fursich

Measure

Item, as you bind on him at the sword, then strike long with the point to his face and wind with the short edge into his face, as depicted here, so you can rush in with pommel or thrust to the face with the armed hand.

Weak

Item, if you bind on him at the sword, then attend to whether he is soft or hard [in the bind]. If he is hard, then wind to his face as described before. If he is soft, then seek the weak of his sword and wind over to your left side, as depicted here, so you can thereafter snap to the head and seek the opening.

Strong

Item, if you have struck above to the sword and he holds strongly against it, then drive up high with your sword behind on his sword against his weak and such that your hilt stands on his sword, and wind your short edge to the neck, as depicted here, so you can pull him to the earth with force.

Before

Item, if you bind another at the sword, and he strikes strongly to your head, then parry with the short edge and press in so he must strike. If he strikes then to the other side then place your sword to his left shoulder, as depicted here, so that you strike into his ear. This is called a Placement and is called Before.

After

Item, if he strikes above to you strongly, then attend and parry his stroke with the short edge. If he is then also quick and strikes before you can lay upon him, then let him strike freely and fall upon his sword with the short edge, as depicted here, thus you bind him and strike him at the ear and go forward.

6r fül*en* ÿnndes

It*em* so dir ain*er* in das sw*er*t pint vn*d* will dir in das an-
gesicht wind*en* oder ander kunst treib*en* so wint auch
auf vn*d* gee reschlich fürsich in in vn*d* als pald er
etwas arbaitt*en* will so uall im starck in sein arm*m*
un*d* stoß in zuruck als hie gm*a*lt stet so wurfstu
in an den ruck*en*

6v It*em* so du aine*m* an pinczt vn*d* er wint dir ein zu
dem gesicht so wint im auch ein vn*d* stee uest in de*m*
winden vn*d* enpfind an seine*m* swert als so dastu
in nit vo*n* deine*m* sw*er*t last köme*n* das er zw
kain*er* arbait mög köme*n* das ist die grösst masß
dez sw*er*cz vn*d* stee starck in der wag als gm*a*lt stet

7r It*em* pincztu aine*m* in das sw*er*t vn*d* er wint dir
ein zu dem gesicht mit sein*er* kurcze*n* sneid so wint
auch hoch auf in dÿ höch vn*d* wen er auf uert vn*d*
wil das v*er*secz*en* so haw im eine*n* u*er*zuckt*en* haw
nach seine*m* elpog*en* alz hie gm*a*lt stet

7v It*em* so du aine*m* nach dem elpog*en* hast gehawe*n*
als da hind*en* gemalt stet v*er*seczt er denn den
haw so stosß im sein sw*er*t nid*er* mit deine*m* gehilcz
vn*d* uall im mit dem knopf vn*d* mit paid*en* arm*m*
v̈ber vn*d* leg im dÿ kurcz sneid ~~nach d~~ an den
hals vn*d* zeuch in als hie gm*a*lt stet

8r It*em* mer ein stuck wen du im nach dem
elpog*en* slechst vn*d* er u*er*seczt denn haw so pleib sten
mit deine*m* sw*er*t an dem seine*n* vn*d* trit für mit
dem recht*en* fuß als welstu im zu der ander*en* seit*en*
slach*en* vn*d* sneid im auß nach seine*m* tenck*en* or*en*
als hie gm*a*lt stet das haist die ausser mÿnn

8v It*em* mer ein sölichs stuck wen dw im nach dem
elpog*en* slechst verseczt er den*n* den slag so uar hoch
auf mit deine*n* paid*en* hend*en* vn*d* lasß dein tencke
ha*n*t uar*en* vn*d* uall im v̈b*er* sein paid arm*m* vn*d* slachs
hint*er* dein tencke seit*en* vn*d* stich mit deine*m* swert
zbisch*en* ~~dir v~~ dein un*d* sein durch vn*d* leg im
das swert an den hals als hie gm*a*lt stet so p*r*ich-
stu im denn arm*m* ab vn*d* sneist im de*n* hals ab

Feeling Meanwhile

Item, if one binds to your sword and wants to wind to your face or perform another technique, then wind up also and go in forward quickly and, as soon as he tries to work something, fall strongly to his arm and thrust him backward, as depicted here, so that he is thrown onto his back.

Item, if you bind on another and he winds to your face, then wind to him also and stand fast in the winding and feel on his sword so that you don't let him come away from your sword and he therefore can't come to any work, which is the greatest measure of the sword, and stand strong in the balance as depicted.

Item, if you bind on another at the sword, and he winds to your face with the short edge, then wind high also, up into the air, and when he drives up and wants to parry then strike a pulled stroke to his elbow, as depicted here.

Item, if you have struck to his elbow, as was previously depicted, and he then parries the stroke, then push his sword down with your hilt and fall upon him with the pommel and with both arms over and place the short edge on the neck and jerk, as depicted here.

Item, another technique. When you strike to his elbow and he then parries the stroke, then remain standing with your sword on his and step forward with the right foot as if you wish to strike to the other side and slice outside to his left ear, as depicted here. This is called the Outer Intention.

Item, another such technique. When you strike to the elbow and he then parries the stroke then drive up high with both hands and let your left go and fall over both his arms and strike behind your left side and thrust with your sword between yours and his through and place the sword at his neck, as depicted here, so that you break the arm and slice away at his neck.

9r Item dir ainer dein swert also hinter sein tencke
ÿgssen geuangen hat vnd wil dich slachen oder stechen
stechen so fleuch hinter in aus seinem slag vnd uall
mit deiner tencken uoren in dein swert vnd wint in
herfür als hie gmalt stet so prichstu im den armm ab

9v Item so dir ainer oben an dein swert pint so wint
im dy kurcz sneid nach dem angesicht ein vnd
gee jnndes vnd trit für mit dem tencken fusß
vnd uall im mit deinem knopf ÿber sein hent vnd
uach dein klingen in dein tencke hant vnd leg
ims an den hals als hie gmalt stet so würfstu
in auf den rucken

Dy geschrift gehört zu dem andrem stuck dauorn

10r Item so dir ainer nach dem elpogen hat geslagen
vnd uächt dir dein swert mit seiner tencken hantt
als uor vnd wil zbischen dir vnd im durch stechen
nach deinem hals als uor so uach sein swercz klingen
in dein tencke hant vnd leg ims selb an seinen
hals als da hie gemalt stet vnd hintertrit in so
würfstu in auf den rucken

Dy geschrift gehört an das hinter stuck vnd iene
gehört her ÿber

10v Item so dir ainer an das swert pintt vnd wint dir
in das angesicht so wint auch vnd lasß den knopf
uaren als uor vnd ual im ÿber vnd hinter trit in vnd
leg im das swert an das haupt als hie gmalt stet
so würfstu in auf den rucken

11r Item wint dir ainer ein in das gesicht so greif
pald mit deiner tencken hant in dein swercz klingen
vnd stich ÿber sein swert im zu seinen hoden d als
hie gmalt stet das ist gar ein gucz namhafcz stuck

11v Item ein gucz swert nemen wen dir ainer oben an
pint an das swert so wint auf mit der kurczen
sneid vnd gee uast inndes in in vnd greif mit deiner
tencken hant in sein pint zbischen seiner hent vnd gee
mit den knopf ÿber sein klingen vnd stoss in gegen
dem maul als hie gmalt stet so nymstu im das swert

Item, [if] one has thus trapped your sword behind his left armpit and wants to strike or thrust, then flee backward away from his stroke and fall with your left [hand] to the front of your sword[5] and wind him forth, as depicted here, so that you break his arm.

Item, if one binds you above at your sword, then wind the short edge to his face and go meanwhile and step forward with the left foot and fall upon him with your pommel over his hands and grab your blade in your left hand and place it at the neck, as depicted here, so that you throw him onto the back.

(This description belongs to the other technique.)

Item, if one has struck toward your elbow and catches your sword with their left hand, as before, and wants to thrust through in between you and himself to your neck, as before, then catch his sword blade in your left hand and place it to his neck, as depicted here, and step behind him so that you throw him onto the back.

(This description belongs to the previous technique and that one belongs over here.)

Item, if one binds you above at your sword, and winds to your face, then wind also and let the pommel drive as before and fall over him and step behind him and place the sword on the head, as depicted here, so that you throw him onto the back.

Item, [if] one winds to your face, then grab quickly with the left hand to your sword's blade and thrust over his sword to his testicles, as depicted here. This is a really renowned, good technique.

Item, a good sword taking. When one binds you at the sword, then wind up with the short edge and meanwhile go strongly to him and grab with your left hand at his grip between his hands and go with the pommel over his blade and punch it to his mouth, as depicted here, so that you take the sword.

[5] That is, to assume the half-sword grip. Thanks to ELIAS FLATSCHER for this observation.

12r It*em* ein anders sw*er*t neme*n* wen dir ain*er* an
pint vn*d* will dir ein wint*en* zu dem gesicht so
greiff pald mit dein*er* tenck*en* ha*n*t in paid swer*cz*
kling*en* vn*d* zeuchs auf dein tencke seit*en* vn*d* gee
mit deine*m* gehilcz vnd*en* an sein hend vn*d* dauch
ṽb*er*sich als hie gm*a*lt stet so ny*m*stu im das sw*er*t

12v It*em* gar ein gucz v*er*porg*en*s stuck für eine*n* iedlich*en*
starck*en* man*n* wen dir ain*er* in das sw*er*t pint vn*d*
wil dir ein wintt*en* od*er* stech*en* so wint auch uast
auf vn*d* gee mit der kurcz*en* sneid starck an sein
swert vn*d* gib im eine*n* gross*en* stosß mit paid*en* hend*en*
fürsich an ~~als~~ so kert er sich uor dir vm*b* so slach
in auf den kopf als hie gm*a*lt stet

13r It*em* mer ain gucz sw*er*t neme*n* wen dir ainer
an pint an das sw*er*t so uar im pald mit dein*er* ~~t~~
rechten hant in sein sw*er*t zbisch*en* sein*er* paider hentt
in das pint vn*d* zeuch an dich vn*d* scheub mit dein*er*
tencken mit dem creüꜩ sein sw*er*t an weg vn*d* stosß in
mit dem knopf in das maul als hie gm*a*lt stet

13v It*em* mer ein sw*er*t neme*n* wenn dir ain*er* an pintt
vn*d* will dir in das gesicht wind*en* so pleib stenn in
der lang*en* sneid vn*d* uar hoch auf vn*d* uar im mit
dem knopf zbisch*en* sein*er* hent ein vn*d*[6] mit der tenck*en*
ha*n*t uasß dein kling*en* vn*d* wint im an den kopff
als hie gm*a*lt stet

14r It*em* mer ein swert neme*n* wenn dir ain*er* an das
sw*er*t pint so pint im auch starck ein vn*d* dauch in hind*er*
sich zu ruck vn*d* lasß dein tencke ha*n*t uar*en* vn*d* greif
im ṽber seine*n* rechte*n* arm*m* in das sw*er*t ~~s~~ czbisch*en* sein*er* he*n*t
vn*d* zeuch hind*er*sich als hie gem*a*lt stet

14v It*em* ein gut stuck wenn dir ain*er* an das sw*er*t
pint so wint im reschlich ein vn*d* ual mit dem knopff
ṽber vn*d* greif mit der tenck*en* ha*n*t in dein kling*en* vn*d*
trit mit dem tenck*en* fuß hind*er* in vn*d* leg ims sw*er*t
an hals vn*d* zeuch in ann ruck*en* vn*d* stichs swert
in in als hie gm*a*lt stet

[6] *vnd* added above the line.

Item, another sword taking. When he binds you and wants to wind to your face, then grab quickly with your left hand to both sword blades and jerk to your left side and go with your hilt under his hand and press upward, as depicted here, so that you take the sword.

Item, a very good forbidden[7] technique for a fairly strong man. When one binds you at the sword and wants to wind or thrust to you, then wind also firmly up and go with the short edge strongly on his sword and give him a great push forward with both hands so that he is turned before you and then strike to his head, as depicted here.

Item, another good sword taking. When one binds you on the sword, then drive quickly with your right hand to the grip of his sword between his hands and jerk toward yourself and shove his sword away with your left [hand] with the cross and punch him with the pommel in the mouth, as depicted here.

Item, another sword taking. When one binds to you and wants to wind to the face, then remain standing at the long edge and drive up high and drive to him with the pommel between his hands and with the left hand grab your blade and wind to his head, as depicted here.

Item, another sword taking. When one binds you on the sword then bind to him also strongly and push him back behind and let your left hand go and grab over his right arm to the sword between his hand and jerk behind, as depicted here.

Item, a good technique. When one binds you on the sword, then wind to him quickly and fall over with the pommel and grab with the left hand to your blade and step with the left foot behind him and place the sword at the neck and jerk him onto his back and thrust the sword into him, as depicted here.

[7] The word *verporgen* can be read as either "forbidden" or "secret", denoting an action that should not be taught to the uninitiated or that might be disallowed in all but the most serious of encounters.

15r Von Ringen dy erst ler

|It*em* zw merck*en* das das Ring*en* will hab*en*
dreÿerlaÿ / sterck / |masß / vn*d* |phentikait / |dÿ sterck
also zw prauch*en* |das ain ÿed*er* nid*er* sol gen in der
wag vn*d* sich starck soll secz*en* auf dÿ erd*en* |dÿ
|masß also das du dein[8] hent vn*d* füsß wist recht
zw schick*en* in all*en* stäntt*en* alstu her nach woll
inne*n* wirst |Dÿ phentikait also dastu woll für
sechst all hint*er* trit zuck*en* vn*d* stoss*en* arm*m* prüch
dastu dar Inn güt gedächtnüsß habst allew
ding pald zw went*en* vn*d* alzeit swär*en* vnd
In dy wag vall*en* **Dÿ ander ler vo*n* Ringen**

|Auch so soltu wiss*en* dastu eine*m* yeden kranck*en*
vor solt ring*en* alz mit sterck |vn*d* eine*m* gleich*en*
mitt solt Ring*en* als mit masß |vn*d* eine*m* starck*en*
nach solt Ring*en* als mit phentikait |Also wen
du mit eine*m* krencker*en* Ringst so darfstu dich
nit vor jm psorg*en* hastu masß vn*d* phentikait vn*d*
seczt dich in ein stäte wag wastu dan suchst
twirch od*er* plösß arm*m* prüch kampfstuck mordstuck
od*er* ander stuck dÿ mag er dir dann nit woll
wer*en* anderst dan ausß fliech*en* vn*d* ausweich*en*

[8] Actually an 'n' with one single dot.

The First Lesson of Wrestling

Item: Remember this, that wrestling needs three things: Strength, Measure, and Agility.

Strength requires that everyone should go low in the balance and should set themself strongly upon the earth.

Measure,[9] so that you know to arrange your hands and feet correctly in all stances, as you will become well aware of , hereafter.

Agility, so that you will set away all steps behind , pulling and shoving, breaking the arm; and that you, therein, have well remembered to avert all things quickly and always make it difficult, and fall into the balance.

The Second Lesson of Wrestling

Also, you should know that you should wrestle each weak opponent Before[10] with Strength, and one who is equal you should wrestle simultaneously[11] with Measure. And one who is strong, you should wrestle After[12] with Agility.

Also, you when you wrestle with an opponent who is weak, then do not be afraid if you have Measure and Agility and set yourself in a balanced stance where you then seek Crossings or openings, arm breaks, fighting techniques, killing techniques, or other techniques that he may not well resist you other than by escaping and dodging.

[9] *Masse* is a quality of chivalric virtue, meaning something like "moderation" or "in equal Moderation". In this use it is related to skilled action or artful application of wrestling skills. See THIEBAUX's analysis of Burkart von Hohenvel's poem "*Min here hat minen sin*" where the qualities *Snel* (swiftness), *wis* (prudence), and *starc* (strong) are applied to a lover, implying that these three qualities have almost a formulaic use in the culture.

[10] "Before" (*Vor*) is a central concept in Liechtenauer's fighting arts, and indicates being the first person to launch an attack that must be responded to by the opponent. The opponent, then, is considered to be in the *Nach* and acting afterwards, and will attempt a parry or block, and work from there to counter the initial action.

[11] Wrestling Simultaneously (*mitringen*) is related to the idea of *Indes* from Liechtenauer's fighting arts, and provides an interesting distinction, yet calls out a relationship, between wrestling and swordsmanship when it comes to actions that happen at the same time. *Mitringen* is one in a list of *Indes* actions given in most glosses of Liechtenauer's couplet about *Indes*.

[12] "After" (*Nach*); see Footnote 10, above.

15v |Ringstu dan mit einem gleichen so sich auch
dastu dich woll pewarst vor zucken hinter treten
vnd armm prüchen vnd vor anderen verfüren das thü
also dastu dich stät in der wag vnd sterck mit
masß vinden last vnd alzeit swärlich in in seczt
vnd in ab arbaiczt dar nach magstu in auß phen-
tikait v̈ber eÿlen mit masß dÿ drit ler etc

|Ringstu mit einem starcken so pewar dich also
wo er dich an greift oder velt so sich dastu vast
nider seiczt vnd ÿm sein armm mit dem peren stosß
auß stost vnd mit anderen stucken auch auß prechst
vnd gar vil vor im weichst vnd siech ob dw in
mit phentikait verfüren vnd v̈bereÿlen mügst
dastu im ein füß erwischst oder sünst vnter
treczt dastu in werfst alstu dann mit mer
stucken vnd pruchen her nach inn gemell vnd
inn geschrift vinden wirst nit mer dan ein vorred

|Doch so ist ein yeder krancker Ringer im ernsten
einem starcken zw gleichen hat er pehentikait
vnd masß kampfstuck vnd mordstuck enpfor
genomen aber mit gesellen ringen so hacz der
starck alzeit enpfor doch so wirt dÿ künst gelopt[13] vor
ritter vnd knechten für allew ding etc daz erst stuck

[13] *gelopt* added above the line.

If you wrestle with one equal to yourself, then also you want to protect yourself from pulling, stepping behind, and arm breaks, and from other enticements. Also that you find yourself standing in the balance with Measure and always set into him heavily and inside from the work, after which you may assail[14] him from Agility with Measure.

The Third Lesson, etc.

When you wrestle with a strong opponent, then beware, also, where he grasps or seizes you. See that you set yourself low and knock out of his arms with the beating strike,[15] and break out with other techniques as well and really give way before him and see if you can mislead him with agility. And you must assail him by sweeping his foot or otherwise stepping under so that you throw him then with more techniques and breaks as you hereafter will find depicted and written.

No more than a forward

Though each weak opponent, who wrestles a strong one in earnest, is equally likely, if they have the Agility and Measure, fighting techniques and killing techniques, to take the advantage, in friendly wrestling however the strong always have an advantage; nevertheless the art is praised as worthy by knights and squires[16] beyond all things, etc.

The First Technique

---

[14] *Uber eilen* is used extensively in this treatise to imply a sense of being "overcome" by the attack. It may be related to the long sword *Hauptstücke* "*überlauffen*" where attacks to higher openings are advocated because they are more likely to hit or require the opponent to react with a parry or block of some kind, rather than launching their own attack. It may also simply mean that one has gone past the opponent's defenses.

[15] The *peren stoss* is referenced in AR on 32v, which says, "If you wrestle with another who is much stronger than you and you cannot break out, then strike his arms off with the beating strike and run in with your head into his belly." This is related to the technique shown in BF 42r. Sadly, this gives us very little technical data other than this technique would appear to target both arms at the same time. *Peren* may also be translated as "bear", though that also gives little clarity.

[16] *Knechten* is a tricky word to translate without more context as it has flexible meanings such as servant, squire, man-at-arms, hired man, or armed servant.

16r |It*em* |So du mit aine*m* ringst zw laufs aus lang*en*
armen / so denck das dein recht*er* arm*m* / auswendig
sey / vnd der tenck in wendig / seczt er dan*n* eine*n*
füß für / so prich auß mit dem arm*m* d*er* geg*en* dem
fuß stee / vn*d* zuck im den füß auf / vnd küm mit
der ander*en* hant / zw hilff / vnd heb in hoch auf /
vn*d* trit mit eine*m* füß / im inwendigs hint*er* seine*n*
füß / vnd tauch in zw ruck |als hy gemalt stet |Das
magstu zw paid*en* seitt*en* treiben

10 stenndt Im Ring*en*[17]

16v It*em* wen er dir den fuß auf wil heb*en* so
zuck den fuß pald hind*er*sich vn*d* reck im starck
von dir vnd stosß in mit der hant ann
das hawpt vn*d* stosß in von dir als hÿe
gemalt stet das get zw paid*en* seitt*en* auch
magstu das sünst mit den hend*en* wer*en*
mit ab stöss*en* vn*d* auß tretten vn*d* wen
du woll in der wag gest nid*er* auf d*er* erd*en*

17r It*em* ist sach das dich ain*er* vb*er* eilt vn*d* dir den*n*
fuß auf zucht so wart dastu den fuß pald
reckst vn*d* sleusß ÿme*n* zwisch*en* sein*er* fusß vnd
halt dich vast an in als hÿe gemalt stett also
magstu woll in dÿ hüff kum*men* vn*d* in den
hacken vn*d* magst in dar aus werffen
das get auch zw paiden seitten vn*d* magst
auch woll ein v*er*porg*en*s stuck treib*en* ob es not tüt

17v It*em* ist sach dastu in denn arme*n* mit aine*m* ringst
so prich auß mit ain*er* hant vnd var im nach dem
fuß alz du in auf welst zuck*en* so zeucht er denn
fuß hind*er*sich vn*d* fleucht da mit[18] so gee dem
fusß nach mit dem arm*m* vn*d* trit mit deinez
fuß hind*er* in in dÿ twirch alz hÿ gemalt stett
das get zw paid*en* seitt*en* vn*d* ÿe nidrer du in
der twirch stest ÿe stercker dw stenn magst
wan dw tarfst anderst nit dann das knie vast pieg*en*

[17] From Paul Hektor Mair's hand.

[18] Damaged word.

Item: When you enter with extended arms to wrestle with someone, then be mindful that your right arm is outside of his and the left inside. As he sets a foot forward, break out with the arm that is on the same side as his foot and pull the foot up and come with the other hand to help and lift him high up. And step with a foot inside behind his foot and press him backwards, as is shown here.

You can do this on both sides.

(10 wrestling positions.)

Item: When he wants to lift your foot up, then jerk the foot immediately behind yourself and extend it strongly away from yourself and strike him with your hand onto his head. And push him away from you as is shown here. You can do this on both sides.

Also, you should defend otherwise with the hands by striking away and stepping out, and when you go well into the balance low to the ground.

Item: In the case when one assails[19] you and pulls your leg upwards, so be sure that you immediately stretch your foot out and lock anyone between his legs and hold yourself against him securely, as is depicted here.

Also be sure you come well into the hip and in the hook and in there make a throw.

Do this on both sides and you should also perform a concealed technique if it is a struggle.

Item: In the event that you wrestle with another at the arms, then break out with one hand and send that in towards the leg, as if you want to pull it up, so that he jerks his foot back and flees with it, then follow the foot with the arm, and step with your foot behind him in the Crossing[20] as is depicted here.

Do this on both sides and the lower you stand in the Crossing, the stronger you will stand when you do this by not changing other than bending the knee.

[19] *Über eilen* carries both the sense of being attacked, but also of being preempted.
[20] The word "*Twirch*" is more familiar to us as the strike *Zwerchhau* or which is sometimes spelled *Twerhau*. This word, "*Twer*" is often used in conjunction with *Krump* ("*krump und twer*") in medieval works to imply something moving "to and fro" or "back and forth" or "crooked and sideways". It can also mean "reversed", "upside down", and "in the way". As "in the way" is accurate, but clumsy in English as a proper noun, I've decided to go

18r It*em* so dir ain*er* ~~in~~ also in dy huff kumpt vnd
wil dich dar auß werffen so denck ~~daﬆu~~ vnd
reck denn fuß vaﬆ hint*er* in vn*d* greiff mit dem
arm*m* der geg*en* im ﬆet ob*en* v̈ber sein agssel vier*en*
fürsich auf dÿ erd*en* als hie gemalt ﬆet so prichﬆu
im denn arm*m* ab auch mag ein ÿeder ringer
sich wol hütt*en* das man*n* im nit in dÿ twirch
kumpt also das er aine*n* von im ﬆoﬆ oder
hinder sich tritt vn*d* sich nit vinden lätt

18v It*em* dÿ and*er* hüf treib also wen dw mit ainem
ringﬆ in denn arme*n* so prich im auß mit dem
recht*en* arme*n* vn*d* var im durch sein tencks
v̈gssen vn*d* tritt mit deine*m* recht*en* füsß im hinter
seine*n* recht*en* fusß alz hie gemalt ﬆet vn*d* halt
in dy weill veﬆ peÿ dem ~~te~~ recht*en* elpogen
vn*d* denck daﬆu dich vaﬆ fürsich senckﬆ am
hin ein spring*en* das er dich nit zw ruck ﬆosß
das treib zw paid*en* seit*en* dar nach vn*d* er ﬆet

dy drit tbirch[21]

19r It*em* so dir ain*er* in dÿ twirch also tritt als er dar
ein kumpt mit dem fuß so ker dein knie geg*en*
im vn*d* tauch in hind*er*sich zw ruck so iﬆ es ge-
proch*en* so dir also geproch*en* iﬆ[22] vnd sichﬆ daﬆu zw
ruck fall*en* müﬆ als paldu dan enpfinﬆ daﬆu
wider hind*er*sich müﬆ so lasß den vodrer*en* arm*m*
resch gen vn*d* slach in hind*er* ÿen vmb das pain
oder auf dÿ erd als hÿ gemalt ﬆet so würfﬆu
in v̈ber dÿ hüff auß das get zw paid*en* seitt*en*

with "crossing" since it pairs better in a sentence with the verb *treib*en—as in "Drive the First Crossing." Please recognize that this isn't perfect, but I think captures the idea most comfortably in English.

[21] Underneath the text, next to the image.

[22] *ist* added above the line.

Item: Then if another comes to you at the hip and wants thereafter to throw you, then anticipate it and quickly extend the foot behind him and, with the arm that is against him, grasp up over his shoulder securely forward toward the earth, as is depicted here, so you break the arm off as well.

Also, make yourself well-guarded against every wrestler so that nobody comes into the Crossing. Also strike him away or step behind and do not let yourself be found.

Item: Drive the second hip, as well, when you wrestle with another at the arms. Break out with your right arm and send it through his left armpit and step in behind his right foot with your right foot, as is depicted here, and meanwhile hold him strongly by the right elbow and be sure that you sink heavily forward as you spring behind him so that he does not strike you backwards.

Drive this on both sides thereafter and [if] he stands.

(The Third Crossing.)[23]

Item: Then if another comes to you in the Crossing, step as he comes to you with the foot, then turn your knee against him and press him backwards to his back, so it is countered.

If you have been countered thusly, and you see that you must fall backwards, as soon as you perceive that you are reversed backwards, then let your forward arm quickly go and strike behind them toward the leg or to the ground, as is depicted here, so you throw him out over the hip.

Do this on both sides.

[23] The drawing on 18v illustrates the text of 19v, which is indicated by this additional text written next to the image.

19v Item die drit twirch treib also wen dw mit aine*m*
in den arm*m* ringst so prich mit aine*m* arm*m* auß
vn*d* gee mit dem selben arm*m* vn*d* fuß durch seine*n*
arm*m* vn*d* halt dich vast fürsich vn*d* halt seinem
arm*m* starck mit paiden henden vn*d* ker dich von
im als hÿe gemalt stett so wurfstu in auff
den rucken das get zw paiden seitten doch so
denck dastu mit eine*m* grossen stosß hinein
treczt das er dich nit zw ruck werffen mög

dy and*er* tbirch[24]

20r Item so dir ain*er* also in die twirch ku*m*pt vn*d* wil dich
v̈b*er* den fuß werffen so secz das knie für oder weich
hint*er*sich mit dem fuß so werst ÿms so dir denn
ain*er* das wertt vn*d* dir die twirch nit lassen wil
so trit pald wid*er* zu ruck vn*d* ny*m* den arm*m*pruch
als hie gemalt stett vn*d* lauf hind*er*sich vmb vn*d*
vmb so magstu in w*er*ffen od*er* du magst wider
in dÿ ~~h~~ twirch treten das get auch zu paid*en* seitten

20v Item dÿ vierd twirch treib also wen dw mit aine*m*
ringst in den arme*n* so prich im auß mit dem
rechten arm*m* vn*d* gee ein weil vmb mit im vn*d* vrb*er*ing
so spring mit dem rechten fusß hinder seine*n* rechten
fuß vn*d* mit dem rechten elpogen hind*er* sein rechs
v̈gsen als hie gemalt stet so würfstu in auf
den rucken das get zw paiden seitten vn*d* müst
dich auch woll fürsich sencken das er dich nit
zw ruck stosß vn*d* dastu in in die wag pringst

[24] Underneath the text, next to the image.

Item: Drive the Third Crossing as well when you are wrestling with another at the arms. Break out with one arm and go with the same arm and foot through his arm and hold yourself firmly forwards. Hold his arms strongly with both hands and turn yourself before him as is depicted here, so you throw him onto the back.

Do this on both sides, however be sure that you step into it with a big shove so that he cannot throw you backwards.

(The Second Crossing.)[25]

Item: Then if another comes to you in the Crossing and will throw you over the foot, then set the knee forward or yield backwards with the foot so it is defended.

If another has defended against you and will not let you into the Crossing, then quickly step again backwards, and take the arm break, as is depicted here, and run backwards around and around, so you can throw him or you can step against him into the Crossing.

Do this on both sides.

Item: Drive the Fourth Crossing as well when you wrestle with another at the arms. Break out with the right arm and go turning around with him a while and spring abruptly with the right foot behind his right foot and with the right elbow past his right armpit, as is depicted here, so you throw him onto the back.

Do this from both sides, and you must also lower yourself well, so that he cannot push you backwards, and that you bring him into the balance.

[25] The drawing on 19v illustrates the text of 18v, which is indicated by this additional text written next to the image.

21r It*em* ein gucz stuck fur eine*n* starck*en* man
wenn dw mit aine*m* an pinczt an das sw*er*t so thü
als welstu im in das angesicht wind*en* vn*d* stosß mit
deine*m* creücz uast an sein sw*er*t vn*d* uar hoch auf vn*d*
lasß dann dein sw*er*t uall*en* vb*er* dein haupt vn*d* uall im
vnd*en* vm*b* paid füß als hie gm*a*lt stet so wurfstu in

21v It*em* ein gucz sw*er*t neme*n* wen*n* dir ain*er* an das
sw*er*t pint so uall im mit dein*er* tencke*n* ha*n*t in paid
sw*er*cz kling*en* vn*d* gee mit dem knopf vn*d* mit der
recht*en* ha*n*t vnt*en* durch sein sw*er*t vn*d* zeuch hind*er*sich
als hie gm*a*lt stet so ny*m*stu im das sw*er*t etc

22r It*em* sticht dir ain*er* mit eine*m* deg*en* ob*en* ein zu dem
gesicht so uar auf mit deiner äbichen tencken
ha*n*t vnd*en* an seine*n* arm vn*d* reib im denn arm*m* vm*b*
vn*d* tu alss welstu in mit deine*m* deg*en* in den hals
stech*en* vn*d* stich in v̈ber seine*n* arm als hie gm*a*lt
stet vn*d* zeuch so p*ri*cht im der arm*m* ab

Im dolchen senndt der stennt
Inn als 14 bar[26]

22v It*em* sticht dir ain*er* ob*en* zu dem gesicht so uass deine*n*
deg*en* auf dein recht*e* ha*n*t vn*d* uar auf da mit vn*d*
uach den stich dar auf vn*d* greiff mit dein*er* tencke*n*
ha*n*t an seine*n* elpog*en* vn*d* heb auf als hie gm*a*lt stet
so p*ri*chstu im den arm*m* vn*d* ny*m*st im den deg*en*

23r It*em* sticht dir ain*er* zu dem gesicht so ny*m* den deg*en*
auf deine*n* recht*en* arm*m* vn*d* uach den stich dar
auf vn*d* wint im mit deine*r* spicz v̈b*er* seine*n* arm*m* vn*d*
zeuch an dich als hie gm*a*lt stet so ny*m*stu im den
deg*en* od*er* du prichst im den arm*m* ab

23v It*em* sticht dir ain*er* ob*en* zu den gesicht so stich mit deine*m*
deg*en* von vnd*en* auf mit abich*er* ha*n*t vmb seine*n* recht*en*
arm*m* vn*d* uar mit dein*er* tencke*n* ha*n*t uor an*n* spicz vn*d*
zeuch nider zu der erd*en* als hie gem*a*lt stet so helcztu
in mit gewalt wildu ledig werd*en* so ny*m* deine*n* deg*en*
in dy tenck ha*n*t vn*d* stich in

26 From Paul Hektor Mair's hand.

Item, a good technique for a strong man. When binding with one on the sword then act as if you wish to wind to his face and push with your cross strongly on his sword and drive up high and let your sword fall over your head and fall below around both his feet, as depicted here, so that you throw him.

Item, a good sword taking. When one binds you on the sword, then fall with your left hand to both sword blades and go with the pommel and the right hand under through his sword and jerk behind, as depicted here, so that you take his sword etc.

Item, if one thrusts above to you with a dagger to the face, then drive up with your inverted left hand under to his arm and wrench the arm around and do as if you wish to thrust to his neck with your dagger and thrust over his arm, as is depicted here, and jerk so that you break the arm.

(Of stances in the bare dagger, there are fourteen.)

Item, if one thrusts above to your face, then hold your dagger on your right hand and drive up with it and catch the thrust thereby and grasp with your left hand to his elbow and lift up, as is depicted here, so that you break the arm and take the dagger.

Item, if one thrusts to your face, then take the dagger to your right arm and catch the thrust thereupon and wind your point under his arm and jerk to yourself, as is depicted here, so that you take the dagger or break his arm .

Item, if one thrusts above to your face, then thrust with your dagger from below up with the inverted hand around his right arm and drive with your left hand before on the point and jerk down to the earth, as is depicted here, so that you hold him with force. If you want to free yourself, then take your dagger in the left hand and thrust to him.

24r Item sticht dir ain*er* oben ein zu dem gesicht so
uasß deine*n* degen in paid hent vn*d* uach denn
stich dar auf vn*d* wint im den knopf über seine*n*
recht*en* arm*m* als hie gmalt stet so slechstu im seine*n*
degen in das angesicht

24v Item sticht dir ain*er* oben ein zu dem gesicht so uasß
deine*n* degen in paid hent als uor vn*d* uach den*n*
scht stich dar auf vn*d* lauf uast hindersich zu ruck
mit im vn*d* dauch ÿene*n* üb*er* seine*n* kopf an denn
hals als hie gemalt stet so zeuchstu in auf dy erd*en*

25r Item sticht dir ain*er* zu dem gesicht so uach denn
stich alz uor vn*d* lasß dein tencke ha*n*t vn*d* fass im
seine*n* recht*en* arm*m* vn*d* reib in vor dir nid*er* gegen
dein*er* recht*en* seiten vn*d* verbürff den degen vn*d* stich
im üb*er* sein agssel vn*d* zeuch in an dich vn*d* reib
im den arm*m* auf den rucken als hie gmalt stet
so pincztu in mit dem degen

25v Item sticht dir ain*er* oben zu dem gesicht so uasß deine*n*
degen auf den recht*en* arm*m* vn*d* uach den stich dar ein
vn*d* stoß in mit der tencken ha*n*t an seine*m* elpogen vo*n*
dÿr das er sich uor dir vmbker vn*d* faß in pey eine*m*
fuß mit deine*m* degen alz hie gemlt stet so würfstu in

26r Item stet ain*er* gegen dir mit aine*m* degen un*d* thu
hast sorg auf in so lauf gar frischlich auf in vn*d*
thu sam welest in[27] oben in das gesicht stechen vn*d*
ker dich uor im vm*m*b auf dein rechte seiten
vn*d* stich in vnd*en* zu den hod*en* als hie gmalt stet

26v Item sticht dir ain*er* oben zu dem gesicht so slach im
mit dein*er* tencken ha*n*t aussen auf sein rechtew
ha*n*t so sticht er sich selb in die hod*en* vn*d* stich in
dann mit deinem degen oben ein zu dem
hals als hie gmalt stet

[27] *in* added above the line.

Item, if one thrusts above to your face, then hold your dagger in both hands and catch the thrust thereupon and wind the pommel over his right arm, as is depicted here, so that you strike his dagger to his face.

Item, if one thrusts above to your face, then hold your dagger in both hands, as before, and catch the thrust thereupon and run fast behind to the back with it and push it over his head to the neck, as is depicted here, so that you jerk him to the earth.

Item, if one thrusts to your face, then catch the thrust, as before, and let go with your left hand and grab his right arm and wrench it forward and down toward your right side, and turn around the dagger and thrust it over his shoulder and jerk him toward yourself and wrench the arm to the back, as is depicted here, so that you bind him with the dagger.

Item, if one thrusts above to your face, then hold your dagger to the right arm and catch the thrust thereupon and push him from you with the left hand at his elbow so that he turns around before you and grasp him by a foot with your dagger, as is depicted here, so that you throw him.

Item, if one stands against you with a dagger and you pretend that you are afraid of him,[28] then run very quickly to him and act as if you wish to thrust to the face and turn yourself around before him to your right side and thrust below to the testicles, as is depicted here.

Item, if one thrusts above to your face, then strike him with your left hand outside to his right hand so that he thrusts himself to the testicles and then thrust above to his neck with your dagger, as is depicted here.

[28] Thanks again to Elias Flatscher for help with this difficult passage.

27r It*em* sticht dir ain*er* vnd*en* zu dem pauch denn
wälsch*en* stich so uasß deine*n* deg*en* in paid hent
vn*d* uall im ob*en* auf seine*n* deg*en* vn*d* uasß mit dein*er*
tenck*en* ha*n*t sein recht*en* vn*d* ker dich reschlich vm*b*
uor im vn*d* zeuch in vb*er* die agssel so prichstu
im den arm*m* ab

27v It*em* sticht dir ain*er* vnd*en* zu dem pauch so uasß dein
deg*en* in paid hent vn*d* uall im auf den stich vn*d*
treib in mit deine*m* degen geg*en* dein*er* recht*en* seit*en*
vn*d* heb vb*er*sich auf in die höch als hie gmalt stet so
winczt u im den deg*en* aus der ha*n*t

28r It*em* sticht dir ain*er* ob*en* ein zu dem gesicht so
uar auf mit dein*er* äbich*en* tench*en* ha*n*t vn*d* prich
im den stich auß vn*d* stich im mit deine*m* deg*en*
zbisch*en* der pain vn*d* heb in auf als hie gmalt stet
so wurfstu in auf den ruck*en*

28v It*em* sticht dir ain*er* mit langem deg*en* nach der
prust als franczoschisch so uar auf mit dein*er* te*n*cke*n*
äbich*en* ha*n*t vnt*en* an seine*n* deg*en* vn*d* leg im deine*n*
deg*en* auf seine*n* rechte*n* arm*m* vn*d* prich im den
deg*en* dar vb*er* auß als hie gm*a*lt stet

29r It*em* vichstu mit aine*m* mit dem messer haut
er dir ob*en* ein so u*er*sechcz mit deine*m* messer
aussen mit der fläch auf den nagel vn*d* gee
pald fürsich mit dem tenck*en* fuß vnd uall im mit
dem knopf vb*er* seine*n* arm*m* als hie gm*a*lt stet so
haustu in in den kopf

Im dusseckhen stennt der stent
Inn alls 8 bar[29]

29v It*em* mer ein gucz stuck haut dir ain*er* zw von
ob*en* nid*er* so uach den slag auf den nagel als uor
vn*d* gee mit dem tencken fuß inndes vn*d* stosß in
mit dein*er* tencken hant sein rechte nider zu
der erd*en* vn*d* haw in vber den kopff als hie g*m*alt stet

[29] From Paul Hektor Mair's hand.

Item, if one thrusts an Italian thrust below to your belly, then hold your dagger in both hands and fall upon his dagger from above and grab his right hand with your left hand and turn yourself quickly around before him and jerk it over the shoulder so that you break his arm.

Item, if one thrusts below to your belly, hold your dagger in both hands and fall upon the thrust and drive to him with your dagger against your right side and lift upward high, as is depicted here, thus you wind the dagger from the hand.

Item, if one thrusts above to your face, then drive up with your inverted left hand and break the thrust away and thrust with your dagger between the legs and lift him up, as is depicted here, so that you throw him onto the back.

Item, if one thrusts with a long dagger to the breast, as the French do, then drive up with your inverted left hand under to his dagger and place your dagger to his right arm and thereby break the dagger away, as is depicted here.

Item: If you fight with another with the Messer and he strikes from above to you, then parry with the Messer outside with the flat on the nail and go quickly forward with the left foot and fall over his arm with the pommel, as is pictured here, so that you strike to the head.

(Of stances in the bare dussack, there are eight.)

Item: Another good technique. If one strikes down to you from above, then catch the stroke on the nail, as before, and go meanwhile with the left foot and push his right hand down to the earth with your left hand and strike him over the head, as is pictured here.

30r Item ein gucz stuck haut dir ain*er* oben zu de*m* kopf
so wint im mit deine*m* messer vnten durch vn*d* uach
den straich dar auf vn*d* uall im mit dem knopf vb*er*
seine*n* arm*m* vn*d* reisß an dich vn*d* trit zu mit dem
tencken fuß vn*d* uar im mit der tencken ha*n*t
vm*m*b den hals als hie gm*a*lt stet so wurfstu
in auf den rucken

30v Item mer ein gucz stuck wen ain*er* oben nid*er*hawt so
wint vnd*en* durch mit deine*m* messer vn*d* lauf im ein
vn*d* uar mit dein*er* tencke*n* ha*n*t hoch vb*er* sein recht*en*
arm*m* vn*d* uasß hint*er* das ügss*en* vn*d* stich mit deine*m*
messer zbischen dir vn*n*d im durch an seine*n* hals
als hie gemalt stet

31r Item haut dir ain*er* oben ein zu dem kopf so v*er*secz
aber mit der fläch vn*d* auf den nagel vn*d* dauch
im sein messer an wege das er slachen müsse vn*d*
wen er den slachen wil so süch die nachent vn*d*
zeuch im den arm*m* ab alz hie gm*a*lt stet

31v Item ein andersß stuck für eine*n* starcken haut dir
ain*er* starck oben ein so uasß dein messer zu gewap-
net*er* ha*n*t vn*d* uach den straich dar ein vn*d* trit
zu im mit deine*m* tencke*n* fuß vn*d* leg im das
messer an den hals als hie gemalt stett

32r Item haut dir ain*er* oben ein zu dem gesicht
so uar auf mit gewapnet*er* ha*n*t vn*d* uach den
slag in das messer vn*d* wint im den*n* knopf
vber seine*n* arm*m* vn*d* reiß hind*er*sich als hie gm*a*lt stet
so prichstu im den arm*m* oder ny*m*st im das messer

32v Item haut dir ain*er* oben ein zu dem kopf so uar
auf mit gewapnet*er* ha*n*t vn*d* uach den straich in
dy klingen als uor vn*d* trit uast hint*er* in mit deine*m*
recht*en* fuß vn*d* leg im den knopf an denn hals
als hie gem*a*lt stet so würfstu in an den ruck*en*

Item: another good technique. If he strikes above to your head, then wind with your Messer under and through and catch the strike thereupon and fall with the pommel over his arm and wrench toward yourself. And step with the left foot and drive with the left hand around the neck, as is pictured here, so that you throw him onto his back.

Item: another good technique. When one strikes down from above, then wind through underneath with your Messer and run in to him, and drive with your left hand high over his right arm and catch him behind the armpit. And thrust with your Messer between yourself and him through to his neck, as is pictured here.

Item: If he strikes above to your head, then parry again with the flat and on the nail and push his Messer away so that he must strike, and when he wants to strike, then seek the oncoming [stroke] and draw the arm away, as is pictured here.

Item: Another technique, for a strong man. If he strikes strongly from above to you, then grasp your Messer in the armed hand and catch the stroke upon it, and step to him with your left foot and place the Messer on the neck.

Item: If he strikes from above to your face, then drive up with the armed hand and catch the stroke on the Messer, and wind the pommel over his arm and wrench backward, as you see pictured here, so that you break the arm or take the Messer.

Item: If he strikes above to the head, then drive up with the armed hand and catch the stroke on the blade, as before, and step fast behind him with your right foot and place the pommel at the neck, as is pictured here, so that you throw him onto his back.

33r Item so dir ain*er* also in dÿ twirch ku*m*pt so denck
dastu dich mit dem knie vast gegen seine*n* fuß
seczt vn*d* wid*er* hind*er*sich treczt so dir dan ain*er* also
dar außget so trit pald wider hind*er*sich vn*d* greif
im mit der rechten hant auff dÿ achssel vnd
lauf mit paiden fussen hind*er*sich vn*d* zeuch in
scheib vmb so würfstu in vn*d* in dem lauf so
magstu wid*er* in dÿ twirch kume*n* das get zu paiden s*eiten*

Der Ringen nach folgen 83 bar stennt[30]

33v Item dy fünft twirch treib also wen du mit aine*m*
ringst in den arme*n* so prich im auß nit[31] dem
rechten arme*n* vn*d* gee ein weil mit im vmb vn*d*
vrbering so zuck in an dich vn*d* gee hoch auf
mit dem rechten arm*m* über seine*n* rechten arm*m*
vn*d* ergreif im seine*n* rechten fuß auswendig
vn*d* hint*er*trit in mit deine*m* rechte*n* fuß seine*n* te*n*cke*n*
vn*d* truck in mit dem knie als hie gemalt
stet so würfstu in das get zw paiden seitten

34r Item so dir ain*er* also mit dem arm*m* üb*er* wil vallen
nach dem füsß so secz dich in dy wag vn*d* dauch
in von dir mit denn arme*n* so er dich dann also
von im stost so greiff im mit der rechten han*n*t
auf seine*n* hals pey dem rechten oren vn*d* zeuch
vast hint*er*sich als hie gemalt stet vn*d* hütt
dich das er dich nit hint*er* tret vn*d* das er dir
mit de*m* haupt nit durchgee das get zu p*aiden* s*eiten*

34v Item dÿ segst twirch treib also wen du mit aine*m*
ringst in den arme*n* so prich im auß mit dem
rechten arme*n* un*d* greif im vrbering an das
recht or un*d* trit mit dem rechten fuß hint*er*
seine*n* tencke*n* fuß als hie gemalt stett so
würfstu in gar hart aber du must im*m*
den rechten arm*m* gar vast halten zu dir vn*d*
tauch in zuruck das get auch zw paiden s*eiten*

30 From Paul Hektor Mair's hand.

31 Should be *mit*.

Item: Then if another comes to you in the Crossing, be mindful that you set your knee strongly against his leg and again step backwards.

When another has stepped out, then step quickly again backwards and grip him with the right hand at the shoulder and run with both feet backwards and circle, pulling him so that you throw him. And in the running[32] you may again come into the Crossing.

Do this on both sides.

(Of bare wrestling there follow 83 stances.)[33]

Item: Drive the Fifth Crossing as well, when you wrestle with another at the arms. Break out with your right arm and go around with him a while and abruptly pull him into yourself and go up high with your right arm over his right arm and seize his right leg on the outside. And step in behind with your right foot behind his leg and press with your knee, as is depicted here, so you throw him.

Do this on both sides.

Item: When he will fall over with his arm after your foot, set yourself in the balance and press him before with the arm.

So then if he also strikes you away, then grip him with the right hand on his head by the right ear and pull firmly behind yourself, as is depicted here.

And watch yourself so that he does not step behind you, and that he does not go through with his head.

Do this on both sides.

Item: Drive the Sixth Crossing as well when you wrestle with another at the arms. Break out with your right arm and grip him abruptly against the right ear and step with the right foot behind his left foot, as is depicted here, so that you throw really hard. However, you must hold his right arm very firmly to yourself and press him.

Do this on both sides.

[32] This phrase may be understood to mean that as you run backwards and circle you can turn back into the Crossing. ELIAS FLATSCHER suggested it could also be less literal and read as "in the process".

[33] My appreciation goes to ELIAS FLATSCHER for clarifying the translation of this phrase.

35r It*em* wen dich ain*er* also gevast hat pey dem
kopf so stoß in mit dein*er* ~~rechten~~ tenck*en*[34] hant an seine*n*
recht*en* elpog*en* so muß er dich lassen so dich ain*er*
von im stost so trit mit dem rechten fuß für
in gegen seine*n* recht*en* füß vn*d* zeuch sein rechte
hant vast in dein tencke seit*en* vn*d* wurff in
vber dein recz pain als hie gemalt stet das
treibstu auch woll zw paiden seitten

35v It*em* dÿ sÿb*en*t vn*d* dÿ lest twirch treib also wenn
du mit aine*m* inn denn arme*n* ringst so prich
im auß mit dem recht*en* arme*n* vn*d* gee vmb mit
im vn*d* zeuch in dan an dich vn*d* secz im dein
rechte hant auss*en* auf sein rechcz knie vn*d* val
im mit der recht*en* agsel in sein recht*en* arm*m*
vn*d* stosß in nider zw der erd als hie gemalt
stet vn*d* hab im denn recht*en* arm*m* auch vast in
dein tencke seit*en* das get auch zw paid*en* s*eiten*

36r It*em* so dir ain*er* das thüt vn*d* dir auf das knie
wil greiff*en* so greif im mit dein*er* tencken hant
hint*er* sein angesicht vn*d* mit dem daum vnt*en*
an dÿ nas*en* vn*d* zeuch vbersich so muß er
dich lassen so er dir das tüt so greif pald mit
dein*er* tencke*n* ha*n*t in sein tencke*n* yns glid vn*d* mit
d*er* rechte*n* in den elpog*en* vn*d* heb in vb*er*sich vn*d* trit
mit dem recht*en* fuß für in als hie gemalt stet
vn*d* würf in das get auch zw paiden seiten

36v It*em* so ain*er* mit dir ringt auß langen armen
so prich im auß mit dem recht*en* arm*m* vn*d* gee
vmb mit im vn*d* dauch inn vast hin vn*d* her
vn*d* vrbering zuck in vast geg*en* dir vn*d* puck
dich vor nider vn*d* var im mit deine*m* haupt
durch sein recht*en* arm*m* vn*d* mit dein*er* recht*en*
hant durch seine*n* recht*en* füß vn*d* trit woll
vnt*er* in als hie gemalt stet vn*d* würf in
vber den ruck auß du müst im wol durch
den arm*m* slieff*en* das get auf paid seitten

[34] *tencken* added above the deletion.

Item: When another has caught you by the head like this, then strike with your left hand onto his right elbow so he has to let go.

If another strikes you away from him, then step with the right foot far in against his right foot and pull his right hand firmly in towards your left side and throw over your right leg, as is depicted here.

You can drive this well also to both sides.

Item: Drive the seventh and last Crossing as well, when you wrestle with another at the arms. Break out with the right arm and go around with him and then pull him in towards yourself and set your right hand to him outside to his right knee and fall in with the right shoulder into his right arm and strike him down to the earth, as is depicted here. And have the right arm also firmly into your left side.

Do this on both sides.

Item: If another does this to you and also will grip the knee, then grip him with your left hand, behind his face,[35] with the thumb below onto the nose and pull him upwards so he must let go.

Then if he does this to you, grip quickly with the left hand at his left forearm and with the right into the elbow and lift up and step with the right foot in front of him, as is depicted here, and throw.

Do this on both sides.

Item: When another wrestles with you from extended arms, then break out with your right arm and go around with him and push him firmly back and forth. And abruptly pull in tight towards yourself and bend over in front of him and drive in through his right arm with the head and with your right hand through his right leg and step in well under him, as is depicted here, and throw him up over the back.

You must slip well through the arm.

Do this on both sides.

[35] *Hinter sein angesicht* is a curious phrase, which I am choosing to translate as “sending his face backwards” as *hinter* can be read as *rückwärts* according to Henning 2014: 152.

37r Item so dich ain*er* also hat gevast so reck denn
fuß hinden weit aushin von dir vn*d* leg dich
oben auf in vn*d* swär dich nider zw der
erden vn*d* hüt dich das er dir kaine*n* fuß
er raichen müg vn*d* halt in auch vast
das er oben nit auch von dir kum vnd
lauf vast mit im hindersich zw ruck als
hie gemalt stet piß das er müd werd
dastu in werffen mögst das get zu paiden s*eiten*

37v Item so sich ainer also auf dich legt vnd sich
hinden vast hinausß spreiczt vn*d* dich hart
druckt so thü als welstu im nach den füssen
greiffen vn*d* nach dem rucken vn*d* swing in
auch hin vn*d* her als vil du magst vn*d* vrbe-
ring so vasß im sein paid hent zw einand*er*
vn*d* halt dÿ starck peÿ den elpogen oder sünst
vn*d* sleuf dan mit dem haupt aus im vn*d* wint
dich hint*er* im vmb als welst an den ruck vallen
vn*d* streck dich als hie gemalt stet das get z*u* p*aiden* s*eiten*

38r Item mer wenn ainer also auf dir ligt vn*d*
wil dich also nider drucken so wart nit lang
vn*d* fasß im sein paid hent vn*d* halt ÿms starck
zw einander vn*d* sicz nider auf den ars als
hie gemalt stet vn*d* würff in v̈ber das haupt
auß vn*d* secz im denn kopf in seine*n* pauch magstu
des nit komen so prich im auß mit eine*m* verporgen
stuck das mag er dir nit wenten dÿ selben
verporgen stuck machtu in allen ringen treiben

38v Item so ainer auf dir ligt vnd sich vast hint*er*
sich swärt dastu nit auf magst vn*d* auch zw
kainem stuck noch ausspruch kumen magst so
vasß in mit einer hant pey eine*m* vinger oder
daum vn*d* prich im denn auß dem glid so dich
ain*er* also vast so greiff im mit deiner ledigen
hant an seine*n* arm*m* voren pey dem glenck
vn*d* ker dich vor im vmb vn*d* zeuch im den arm*m*
v̈ber dein achssel als hie gemalt stet vn*d* prich
im den arm*m* enczway wildus prechen so zeuch hintersich

Item: If another has caught you, then extend your feet backwards, far out from yourself, and lay yourself on top and down heavily to the earth. And guard yourself so that he cannot possibly reach your foot and hold strongly as well so that he can not get away from you. And keep[36] running backward behind oneself with him, as is depicted here, until he gets tired, then you shall make the throw.

Do this on both sides.

Item: So, another is laying on you, and is sprawled out strongly backward and presses you hard, then do as if you wish to catch him behind the foot and by the back and swing him thus, back and forth as much as you like. And abruptly catch both of his hands to each other and hold them strong by the elbow or otherwise and then slip from him with the head and wind yourself around behind him as if you wish to fall onto the back, and straighten yourself, as is depicted here.

Do this on both sides.

Item: Further, when another lays on you and also will press you down, then do not wait long and catch both of his hands and hold them strongly together and sit down onto your buttocks, as is depicted here, and throw him up over the head, and set the head into his stomach.

If you do not come to this, then break him out with a hidden technique so that he may not reverse it. You may perform the same hidden techniques in all wrestling.

Item: Then, if another lays on you and holds himself backwards so heavily that you cannot get up, and you cannot come to any techniques, nor break out, then catch him with a hand by a finger or thumb and break it from the joint.

If someone holds you thus, then grasp him with your free hand on the forearm by the wrist and turn yourself around before him and pull his arm over your shoulder, as is depicted here, and break his arm in half.

If you want to counter, then pull backwards.

[36] Thank you to ELIAS FLATSCHER for the insight that "*fest*" can be read as doing something continuously.

39r It*em* so ain*er* mit dir ringt auß langen arm*m* so p*ri*ch
auß mit der recht*en* hant vn*d* gee ein weil vmb mit
im vrbering so var im mit deine*m* haupt durch
seine*n* arm*m* vn*d* ker dich woll vor im vmb als hÿe
gemalt stet vn*d* würff in v̈ber das haupt außß
vnd thü das pehentiklich das er dich nit zu ruck
ziech vn*d* dich nit hint*er*tret das get zu paiden s*eiten*

39v It*em* So du mit aine*m* ringst in langen arme*n* so
prich im auß mit deine*m* rechten arme*n* vn*d* gee
ein weil vmb mit im vn*d* zeuch in starck hin
vn*d* her vn*d* vrbering so ker dich vor im vmb
vn*d* zeuch im denn rechten arm*m* auß dein rechte
agssel vn*d* greiff zw mit deine*r* recht*en* hant
im vor*en* in das glenck als hie gemalt stett
vn*d* würff in also das get auf paid*en* seitten

40r It*em* |So dich ain*er* also gevast hat als in den voder*en*
zwaÿ stucken gemalt stet vn*d* als pald er vor dir
sich vmb will ker*en* vn*d* dir den arm*m* auf dy agssel
pringt so secz dich starck in dy wag vn*d* greiff
im mit deiner tencken hant hinden zwischen
seiner pain vn*d* heb in auf als hie gemalt stett
so wurfstu in fürsich auf das angesicht oder stosß
in mit eine*m* fuß in dÿ kniepüg das get auch z*u* p*aiden* s*eiten*

40v It*em* so du mit aine*m* ringst in langen arme*n* so prich
im auß mit deine*m* recht*en* arm*m* vnd var zw stund
damit an seine*n* rechten elpogen vn*d* fasß im denn
arm*m* starck in dein paid hent vn*d* schlaipf dÿ hent
paid her*r* für an seine*m* arm*m* pisß an das glenck
vn*d* zuck in starck nach dir vn*d* ker dan dein te*n*cke
seitt*en* gegen im an sein recht*en* elpog*en* als hÿe
gemalt stet vn*d* prich im de*n* arm das get z*u* p*aiden* s*eiten*

41r It*em* mer ein stuck wen dw mit aine*m* zw laufs wild
ringen vnd gar ein starcker ist so greif in kecklich an
als welstu mit gar grosser sterck an in vn*d* als pald er
dich mit gwalt zw ruck daucht so secz im eine*n* fusß
auf den nabel vn*d* val pald nider auf den ars vnd
halt deinew knie nahot zw einander als hie gemalt
stet vnd würf in v̈ber dich auß vn*d* halt in starck pey
denn hend*en* so müß er auf das angesicht vallen
das magstu mit paiden fussen thun vn*d* piß snell

Item: If another is wrestling with you from extended arms, then break out with the right hand and go around with him a while, then abruptly drive in through his arm with your head and turn yourself well around before him, as is depicted here, and throw him out over the head. And do this nimbly so that he cannot pull you back and does not step behind you.

Do this on both sides.

Item: If you wrestle with another at extended arms, then break out with your right arm and go around a while with him and pull him strongly back and forth and then abruptly turn yourself around before him and pull his right arm out over your right shoulder and grasp his forearm at the wrist with your right hand, as is depicted here, and thus throw him.

Also do this on both sides.

Item: If another has grasped you as in the two techniques previously depicted, and as soon as he wants to turn around before you and brings your arm outside the shoulder, then set yourself strongly into the balance and grasp him with your left hand from behind between his legs and lift him up, as depicted here, so that you throw him before you to his face.

Or hit him with a foot into the knee.

Do this on both sides.

Item: Then if you wrestle with another at extended arms, break him out with your right arm and place it immediately[37] on his right elbow and grasp his arm strongly in both of your hands, and slip both hands in front of his on his arm to the wrist and pull in strongly to yourself, and turn your left side to him on his right elbow, as his depicted here, and break his arm.

Do this on both sides.

Item: One more technique when you will enter to wrestle with another, and he is much stronger, so grasp onto him boldly, as if you wish to rush him with much greater strength, and as soon as he presses you back powerfully, then set a foot onto the navel and fall quickly back to your buttocks and hold your knees close to one another, as is depicted here, and throw him out over yourself, and hold him strongly by the hands so he must fall onto the face. You may do this with both feet, and be quick.

[37] Thank you to Elias Flatscher for this insight.

41v It*em* ringstu mit ainem gleichs fass*en*s in den arme*n*
als dÿ pawr*en* thund so senck dich vast nider in dy wag
vnd sleuß dein arm*m* nit zw¨ vmb in dastu sy prauchen
mügst vn*d* wen dich ainer dann mit gewalt heben
will so wart als pald er sich nider senckt nach der
krafft vn*d* dÿ knie pew¨gt so stosß in mit deiner knie
ainem aussen an sein knie als hie gemalt stet so
würfstu in auf den ruck das zw paiden seitten

42r It*em* ringstu zu laufs mit aine*m* so thü als welstu
in peÿ dem kopf vach*en* vn*d* reck dich hoch auff
so vert er auch auf mit denn arme*n* als ers wer*en*
well alstu in nü also in dy höch v*er*fürt hast So
puck dich resch vor im nÿder vn*d* stosß in mit dem
kopf vor*en* in dÿe prust oder auf den pauch vnd
var im mit paiden henden vnden vmb dÿ fuß als
hie gemalt stet vn*d* würf in an ruck das get snel zw

42v It*em* mer ain stuck wen dw mit aine*m* zw laufs
wild ringen so wart wen er zw dir kumpt vnd
vn*d* wil dich mit denn arme*n* an vallen so stosß im
mit deine*m* recht*en* fuß nach seine*m* tencke*n* knie vnd
mit der recht*en* hant slach im nach dem tencken or*en*
so zuckt er den fuß vn*d* das haupt hind*er*sich so slach
in mit deine*m* tencken fusß an seine*n* recht*en* als hÿe
gemalt stet so velt er pistu anderst snel daz g*et zu* p*aiden* s*eiten*

43r It*em* so du aine*n* heb*en* wild vn*d* er stost dich mit seinez
knie aussen an dein knie dastu zw ruck müst vall*en*
so reck dein paid fuß von dir[38] ob der erd hin neben sein
vn*d* val mit der hant gegen der erd*en* vast hinder
seine*n* ruck vn*d* halt dein haupt auch hinder sich
als hie gemalt stet vn*d* halt in mit der ander*en*
hant vest so wurfstu in oben v̈ber dich ausß
vn*d* wen*n* er dich zw ruck würft so ny*m* dir ein güte*n*
swunck so muß er her v̈b*er* das get zw paid*en* seit*en*

[38] *dir* added above the line.

Item: If you wrestle with another at the arms with an equal grip, as the peasants do, then be sure that you are low in the balance and do not lock your arms around, in that you may need them. And when another wants to lift you powerfully, then pay attention and as soon as he sinks low to power this, and the knees bend, then strike in with your knee outside onto his knee, as is depicted here, so that you throw him to his back.

Do this on both sides.

[39]Item: If you wrestle with another from the entry, then act as if you will catch him by the head and stretch yourself towards him high so that he also tracks you upwards with the arms, as he wants to defend, so you now also have tricked him high. So duck swiftly downward before him and strike forward with the head into the breast or stomach, and drive in with both hands below around his legs, as is depicted here, and throw him onto the back. This goes quickly.

Item: One more technique when you will wrestle with another from the entry. Watch for when he comes to you and will fall onto you with the arms, then strike him with your right foot behind his left knee, and strike him to the left ear with your right hand so he pulls the foot and the head backwards, then strike in with your left foot on his right, as is depicted here, so he falls if you are quick the second time.

Do this on both sides.

Item: When you will lift another[40] and he strikes with his knee outside on your knee that you must fall backwards, then stretch both feet away from yourself to the earth near his [feet], and fall with the hand going to the earth, far behind his back[41] and also hold your head backwards, as is depicted here, and hold with the other hand fast so you throw him up over yourself.

And when he will throw you backwards, then take a good swing so he must go over you. Do this on both sides.[42]

[39] According to Reiner Welle, folia 41/48 and 42/47 were switched. I concur with this observation, which solves organizational problems in the way it is currently bound. See Welle 2014: 52.

[40] As in the previous technique, 41v.

[41] This instruction is confusing but indicates that you are throwing your hand to the ground behind the opponent's position.

[42] This could be read as a reiteration of the instructions of this counter, or as an instruction that if you're being thrown, roll with it so you flip them and end up on top.

43v It*em* mer ein ſtuck gleichs fass*en*s in den arm*m*
ob du mit eine*m* ſtarcken ringſt vn*d* ob er dich v̈ber
eilt oder wie es sich schickt das er dich auf hebt
v̈ber die erd*en* vn*d* wil dich werffen so greiff im
mit dein*er* tencken hant hint*er* seinez kintpacken vn*d*
tauch in ſtarck von dir als pald er dan die arm*m*
auf let gen so greiff in dein rechte hant in das
glenck als hie gemalt ſtet vn*d* trit mit dem recht*en*
fuß für in so wurfſtu in vn*d* prigſt im dem arm*m*
das get zw paid*en* seitt*en* vn*d* iſt gut für ſtarck leüt

44r It*em* wen dich ainer also gevaſt hat vn*d* paid
arm*m* in ein ander geschlossen hat vn*d* mit dem
füß für wil tret*en* so greiff pald mit dein*er* ~~tencken~~
recht*en*[43] hant in dein ~~rechte~~ tencke[44] hant vn*d* zewch also mit
paid*en* hend*en* herfür wercz als hie gemalt ſtet
so prichſtu im denn arm*m* selb enczwaÿ vn*d* iſt das
ſtuck da mit geproch*en* vn*d* get zw paid*en* seitt*en*
vn*d* iſt ein gucz ſtuck vn*d* pruch vn*d* nüczt es in vil ſtuck*en*

44v It*em* ringſtu mit aine*m* in den arme*n* gleichs fass*en*s
iſt dir ain*er* zw ſtarck vn*d* vn*d* hebt dich mit gwalt
auf vn*d* wil dich werffen so var im mit dein*er* tencke*n*
hant hint*er* den kinpacken vn*d* mit der recht*en*
zwisch*en* sein*er* pain vnd fasß im das recht pain
vn*d* heb in auf vn*d* tauch in ob*en* hin dan von dir
als hie gemalt ſtet das get zw paid*en* seitt*en*

45r It*em* wil dich ain*er* also vassen wil[45] hint*er* dem kinpack*en*
so hab das haupt vn*d* angesicht nachot an in
auf sein prüſt das er nit dar vnt*er* müg so dir
ain*er* das werd so greiff im aussen vmb das
haupt auf dÿ nasen oder in das maul als hie
gemalt ſtet vn*d* würf in v̈b*er* ein pain vn*d* hutt
dich das er dich nit peisß das get zu paid*en* *seiten*

[43] *rechten* added to the left of the text area.
[44] *tencke* added above the deletion.
[45] *wil* added above the line.

Item: One more technique from the equal grip at the arms. If you wrestle with one who is stronger and if he assails you or as he intends to lift you up over the earth and wants to throw you, then grip him with your left hand, behind his jaw, and press him strongly from you. As soon as he lets go of the arms, then grip with your right hand into the joint, as is depicted here, and step in front of him with the right foot, so you throw him and break his arm.

Do this on both sides, and it is good for strong people.

Item: When another has thus gripped you, and has both arms locked and wants to step with the foot before you, then grasp quickly with your right hand in your left hand and pull as well with both hands down forward, as is depicted here, so you break the same arm in half.

And the technique is thereby countered, and do it on both sides, and it is a good technique and counter and can be used in many techniques.

Item: When you wrestle with another in the equal grip at the arms, and they are too strong and lifts you up with force and wants to throw you, then drive him with your left hand behind the jaw, and with your right between his thighs and grip him by his right thigh and lift up, and then press him over backwards from yourself, as is depicted here.

Do this on both sides.

Item: When another intends to grip you behind the jaw, then have the head and face near in towards his breast so that he cannot get underneath.

Then when another defends like this, grab him outside around the head to the nose or in the mouth, as is depicted here, and throw him over a thigh and guard yourself so that he does not bite you.

Do this on both sides.[46]

[46] It is interesting that the artist drew an additional hand on the far shoulder, perhaps showing a "friendly" version of this counter where one isn't relying upon cranking the neck to power the throw. CHRISTIAN TOBLER also suggested this as a way to avoid being bitten yet to still achieve a throw from the same opening.

45v It*em* wen du mit aine*m* ringſt gleichs vassens so ge-
denck wen dich ein ſtarcker zw im trucken will so wal
im mit deine*n* paiden arm*m* vmb seine*n* rechten arm*m*
als hÿe gemalt ſtet vn*d* secz im dein rechte achssel
ſtarck hint*er* sein rechte achsel so machſtu in müed
vn*d* siech dÿ weil ob dw eine*n* fortail vind*en* mügſt
vn*d* ob du im ein fuß mügſt neme*n* das get zw p*aiden seiten*

46r It*em* mer ein ringen wen du mit aine*m* in den
arme*n* ringſt gleichs vass*en*s iſt er dir zw ſtarck
vn*d* magſt dich sein nit wol wer*en* so reck dich
weit hinden hinauß vn*d* faß in mit dein*er* recht*en*
hant in sein*er* tencke*n* seit*en* mit der tencke*n* auf sein
rechte hüf vn*d* wen dich tunckt das er vaſt für-
sich dring so greif im mit der tencke*n* hant hind*en*
zwisch*en* der pain vn*d* heb in auf als hie gemalt
ſtet so velt er das get auch zw paid*en* seitten

46v It*em* wen du mit aine*m* ringſt mit gleichem vassen
in den arme*n* so dir ain*er* denn zu ſtarck iſt vn*d* will
dich zw im druck*en* so lasß dy tenck hant pald var*en*
vn*d* ker dein rechte seitt*en* gegen im vn*d* trit mit de*m*
recht*en* fuß zwischen seiner füß als hie gemalt ſtet
dar nach magſtu ringen auß der hüf oder auß
dem hacken alſtu dan vor vnt*er*weiſt piſt word*en*
das iſt ein ſtuck da vaſt alle ringen anligen
vn*d* iſt nit pillich das ma*n* si mal oder schreib

Item: When you wrestle with another in the equal grip, then remember when he is stronger and wants to pull you in strongly, then fall in with both arms around his right arm, as is depicted here, and set and set your right shoulder strongly behind[47] his right shoulder so you make him weary. And see if you can find an advantage or if you can take a foot.[48]

Do this on both sides.

Item: One more wrestling when you wrestle with another at the equal grip. If he is too strong for you, and you cannot defend yourself well from him, then stretch out wide far backwards and grip him with your right hand on his left side, with your left to his right hip. And when you think that he almost gets through, then grip him, with your left hand behind, between his legs[49] and lift it up, as is depicted here, so that he falls.

Do this on both sides.

Item: When you wrestle with another with equal grips at the arms, so that another is stronger than you and wants to pull you in to him, so let your left hand quickly drive and turn your turn your right side toward him and step with your right foot between his feet, as is depicted here, thereafter you may wrestle from the hip or from the Hook, as you have been instructed.[50]

This is a technique that sets up nearly all wrestling and it is not appropriate that it is shown or written.

[47] This is another case where the word *hinter* may indicate a direction of movement rather than a placement in relative space. See footnote 35.

[48] This appears to refer to the following technique (46r).

[49] This is a curious instruction, as there is no indication of the instruction to go between in the image. One option is that we are driving our left hand through and between, which would engage their leg deeper than is depicted in the image. The other possibility is that we should be bringing our right hand between the legs, but that the scribe forgot to indicate the right-hand part of the instruction. This could make the throw similar to Kal's 4th wrestling at the hip (PKM, f 85v).

[50] Thank you to Dr. FLATSCHER for providing clarity regarding the meaning of this final phrase.

47r It*em* mer ein stuck wen dir ain*er* denn hacken also
ein hat geslag*en* vn*d* sich vast in dich spreczt so
dring in ein weil von dir vn*d* ṽrbering so heb in
auf vn*d* slach in mit den paiden geslossen fussen
an seine*n* ledigen fuß als hie gemalt stet so müß
er zu ruck vall*en* das get zw paiden seitten
vnd ist nit woll zw went*en* dan ain*er* der gar wol
heckl*en* kan*n* der hüet sich woll dar vor ab*er* selt*en*

47v It*em* so dir ain*er* den hacken ein sleüst vn*d* vast dich
pey dein*er* ledig*en* hent vn*d* senckt sich gen der erd*en*
vn*d* wil dich auß dem hacken werffen so greif mit
der ander*en* hant ob*en* ṽber sein achssel vn*d* fasß in
peÿ der nasen als hie gemalt stet so prichstu im
auß das er dich auß dem hacken lassen müsß
dar nach such eine*n* vortail wie du mügst pey
eine*m* fuß oder mit hint*er*tret*en* das zw paiden seitt*en*

48r It*em* ringstu mit aine*m* in den arme*n* gleichs vass*en*s
so dir ain*er* dan zw starck ist vn*d* wil dich mit ge-
walt zw im rucken als paldu des enpfinst so zuck
den recht*en* arm*m* auß im vn*d* val im mit dem tencke*n*
arm*m* ṽber sein rechte achsel vn*d* ker dein tencke*n*
seitt*en* vast gegen im vn*d* slach im dein tencken
fuß ein in den hacken als hie gemalt stet so
magstu aber alle hüfringen vn*d* hinter tret*en* als
vor geschriben stet das gat als zu paid*en* seit*en*

48v It*em* wen du mit aine*m* also ringst als in dem ander*en*
stuck vor geschrib*en* stet vn*d* als pald er dir denn
hacken vmb den fuß slecht so heb den selbigen
fuß auf vnd trit im auf seine*n* ledigen fuß als
hie gemalt stet vn*d* puck dich pald vor*en* nider
vn*d* stosß in von dir vn*d* zuck im das haupt ausß
dein[51] ṽgssen das get zw paid*en* seitten etc

[51] Might be *dem*.

Item: One more technique when another has struck into the Hook against you, and he wedges firmly into you, then push him from yourself a while, and suddenly lift him up and hit with the feet locked together onto his free foot, as is depicted here, so he must fall to his back.

Do this on both sides, and it is not good to reverse, even those very good at Hooking can rarely then protect themselves well from this.

Item: So another locks you in the Hook and holds you by your free hand and lowers himself to the earth and wants to throw you from the Hook, so grip with your other hand up over his shoulder and take him by the nose, as is depicted here, so you break him out, that he must let you go from the Hook.

Thereafter search for an advantage however you can by a foot or with the stepping behind.

Do this on both sides.

Item: When you wrestle with another from the equal grip at the arms, if another is too strong and wants to violently pull you to him, as soon as you feel this, so pull your right arm out from him and fall to him with your left arm over his right shoulder and turn your left side nearly to him and hit with your left foot into the Hook, as is depicted here.

You may, however, come to all hip wrestling and stepping behind as described and written before.

Do this on both sides.

Item: When you wrestle with another in the other technique described before, and as soon as he strikes around your foot into the Hook, then lift the same foot up and step into his free foot, as is depicted here, and duck down forwards and push him from yourself, and pull his head from your armpit.

Do this on both sides.

49r It*em* mer ein stuck auß dem hacken wen er dir
den hacken ein hat geslagen vn*d* deine*n* ledigen
arm*m* gevast hat als vor vn*d* wil dich werffen als
er sich reckt gegen der erd*en* so lasß den hinter*en*
arm*m* v̈ber sein achsel gen als hie gemalt stet
vn*d* reib dich mit der hüf für in vn*d* greiff mit dem
ob*eren* arm*m* nider gegen der erd*en* so prichstu im
seinen arm*m* enczwaÿ vn*d* würfst in das g*et* z*u* p*aiden* s*eiten*

49v It*em* so dw aine*m* den hacken ein hast geslagen vn*d* du
enpfinst das er dich dar auß wil pringen so spann
dich starck in den hacken vn*d* swär dich auf in vnd
vrbering so trit mit dem selbig*en* fuß für in vnd
ker im deine*n* ruck gar in sein schosß als hie gemalt
stet vn*d* würff in v̈ber den ruck aus vn*d* halt dich
starck fürsich das er dich nit zu ruck ziech vn*d*
greiff mit der aine*n* hand hind*en* zw dem pain
so magstu in dest*er* leicht*er* heb*en* das get zw p*aiden* s*eiten*

50r It*em* so dich ain*er* also hat gevast vn*d* kumpt dir mit
dem ruck für den pauch als paldu siechst das er
den trit wil nemen mit dem recht*en* füß so greif
im resch mit dein*er* recht*en* hant hint*er* sein rechcz pain
vn*d* heb in auf vn*d* würf dann als hie gemalt
stet vn*d* wen du in nün auf hast ghebt so greif
mit deiner tencken hant im auf sein hals pey
dem recht*en* or*en* so magstu in w*er*ff*en* das get z*u* p*aiden* s*eiten*

50v It*em* so dich ain*er* also pey dem fuß gevast hat vn*d* hebt
dich auff vn*d* wil dich auf den kopf werffen vnd in
dem als er im denn sbunck nÿ*m*pt vn*d* dich gegen der
erd wil stossen so swing dich mit sampt seine*m* sbu*n*g
vn*d* slach im deine*n* tencken arm*m* hind*en* vn*d*[52] sein te*n*cks
pain vn*d* fasß in mit der recht*en* hant ob*en* vmb den
hals vn*d* heb in dan auf als hie gemalt stet vn*d*
sleuß dan dein paid hent in ein ander vn*d* lauf
vm*b* mit im das er tämisch werd so würfstu in
an mü vn*d* an schad*en* das get auch auf paid seit*en*

[52] Should be read as *vmb.*

Item: One more technique out of the Hook. When he has struck you into the Hook and has ahold of your free arm, as before, and wants to throw you, as he wrenches you to the earth, so let the arm that is behind go over his shoulder, as is depicted here, and slide your hip far in front and with the arm that is over, grasp down towards the ground, so you break his arm in half and throw him.

Do this on both sides.

Item: If another has struck you into the Hook and as soon as you realize that he wants to bring you out, then lash yourself strongly and heavily into the Hook, and then unexpectedly step with the same foot far forward and turn into him completely with your back into his lap, and is depicted here. And throw him over the back and hold yourself strong in front of him so that he does not pull you back, and grip with your other hand behind to the leg so you may more easily lift him.

Do this on both sides.

Item: If another has grabbed you like this and comes to you with the back in front of your stomach, as soon as you see that he then will take the step with the right foot, then grip him swiftly with your right hand behind his right thigh and lift up, and then throw, as is depicted here. And when you have lifted up, then grip with your left hand into his head by the right ear so you may throw him.

Do this on both sides.

Item: If another has gripped you by the foot lifted you up and wants to throw you to the head and in this as soon as he swings you, and you will be struck to the earth, so swing yourself with the direction of his swing and hit him around his left leg with your left arm and grip him with your right hand over around the head and, in that, lift up, as is depicted here. And then lock both hands in together and run around with him in this so that he becomes dizzy, so you throw him without trouble and without harm.

Do this also on both sides.

51r It*em* so du mit aine*m* ringst vn*d* er v̈ber eilt dich mit
sterck oder phentikait das er paid arm*m* vnt*er* pringt
vn*d* wil dich auf heb*en* so senck dich in die wag vn*d*
reck im deine*n* rechte*n* fuß pald zwischen sein*er* pain
vn*d* vasß in mit dem recht*en* arm v̈b*er* sein tencke achssel
pey dem rechten v̈gsen als hie gemalt stet so magstu
alle die stuck treib*en* die aus dem hacken vn*d* aus
der huf gend vn*d* magst das auf paid*en* seit*en* treib*en*

51v It*em* mer ein stuck so dich ain*er* v̈ber eilt vn*d* paid
arm*m* vnt*er*pringt so secz dich fast nid*er* in dÿ wag vn*d*
hüt dich wol das er dich nit hint*er*tret so müß er dich
mit gewalt auf heb*en* wil er dich anderst werffen
vn*d* als pald er dich gehept vn*d* werffen wil so greif
im mit dein*er* tencken hant hint*er* dy augen vn*d* mit
dem daum vasß in pey der nas*en* vn*d* dauch in v̈b*er*sich
als hie gemalt stet das magst mit paid*en* hend*en* thün

52r It*em* wen du aine*n* also auf gehebt hast als vor stet
gemalt vn*d* er greift dir nach der nasen vn*d* wil
dich weg dauchen so halt in mit deiner tencken
hant vest pey seine*m* ruck vn*d* secz deine*n* tencken
fuß mit dem knie aussen an sein knie vn*d* greif
im mit deiner rechten hant an seine*n* tencken
elpogen als hie gemalt stet vn*d* stosß in auf in dÿ
höch vn*d* würf in von dir das get von paid*en* seitten

52v It*em* so dich ain*er* alsod hat pey dem elpogen vn*d* dich
ob*en* hinv̈ber daucht als dan hinden gemalt
stet so zuck deine*n* recht*en* fuß hindersich von
seine*m* knie vn*d* greiff mit deiner tencken hant
nach seine*m* tencken füß vn*d* trit mit deine*m* tencke*n*
hint*er* seine*n* rechten vn*d* dauch in mit dem knie
vn*d* heb in auf als hie gemalt stet so würfstu
in auf den rucken vn*d* küm dan der ~~rechten~~
tencke*n* ha*n*t zu hilf mit der rechte*n* das get *zu paiden seiten*

Item: When you wrestle with another and he assails you with strength or agility so that he brings both arms under and wants to lift you up, then sink yourself into the balance and bar him with your right foot quickly between his legs and grab him with the right arm over his left shoulder by the right armpit, as is depicted here, so you may drive all techniques from the Hook and execute them from the hip.

And you may do this on both sides.

Item: One more technique. If another assails you and brings both arms under, then set yourself firmly down in the balance, and guard yourself well so that he does not step behind you, thus he must lift you with force. If he wants to throw you a different way, as soon as he lifts and intends to throw you, then grip him with your left hand, sending the eyes backwards,[53] and grab him by the nose with the thumb and press him upward, as is depicted here.

You may do this with both hands.

Item: When you have lifted another up, as is described before, and he grabs your nose and intends to press you away, then hold him firmly with your left hand at his back and set your left foot out with your knee on his knee and grip with your right hand on his left elbow, as is depicted here, and push it up high and throw him from you.

Do this on both sides.

Item: If another has you by the elbow and presses you up over backwards, as previously described, then pull your right foot behind yourself away from his knee and grip with your left[54] hand to his left foot and step with your left behind his right, and press in with the knee and lift up, as is depicted here, so you throw him on the back. And then come to help your left hand with your right hand.

Do this on both sides.

[53] I decided to translate this phrase according to the conventions I have previously used, but this could also be read something like "make him look backwards". Dr. FLATSCHER suggested "hinter" in this case could indicate "over the eyes".

[54] Read: "right".

53r It*em* so du mit aine*m* ringst vn*d* hat aber paid
arm*m* vnden vn*d* ist gar ein starcker man vn*d* hept
dich auf vn*d* wil dich mit gbalt auf dÿ erd stossen
oder werffen oder mit dir vm*b* lauff*en* oder
wil dich hart drucke*n* als paldu v*er*nÿ*m*st das er
dir einenn schad*en* wil thün so greif im resch
vn*d* starck mit paiden hend*en* auf sein haupt vn*d*
dauch ims ein wenig zw ruck vnd druck
nider als hie gemalt stet so stostu im das haupt
vn*d* den hals in den potig das er nit red*en* mag

53v It*em* so dich aber ainer also ÿbereilt vn*d* paid arm*m*
vn*d* paid arm*m* vnt*er*pringt so sich dastu dich vast
nider senckst vn*d* auß im winczt vn*d* wen er dich
dan zw im zucken wil vn*d* dich auf wil heben
so vmb schleuß im sein paid arm*m* mit deine*n* paid*en*
hend*en* vn*d* druck vast zam*men* vn*d* secz im dein paid
few¨st in seinen magen als hie gemalt stet
vn*d* tauch in also von dir vn*d* wart ob du in vnt*er*
treten mügst oder ob du in dÿ huf küme*n* mügst

54r It*em* so dich aber ain*er* ÿb*er* eilt vn*d* paid arm*m* vnt*er*
pringt so halt dich mit paiden arm*m* ob*en* vmb
den hals vest zwn im vn*d* slach im deine*n* rechten
fuß auch vmb seine*n* tencke*n* aussen her ein vn*d*
greiff den mit ainer hant welches ist aussen
vmb sein haupt vn*d* fasß in pey der nasen als
hie gemalt stet vn*d* zeuch in hint*er*sich da mit
vn*d* secz im eine*n* fuß für so würfstu in auf de*n* ruck

54v It*em* ob ain*er* mit dir ringt vast dich mit seine*n* paid*en*
hend*en* peÿ der joppen als auf den hüffen vn*d*
slingt dich hin vn*d* her vn*d* helt dich so starck
das tu ny*n*dert zün im kumen magst so greif
im ein weil in sein elpog*en* ÿnbendigs vn*d* dauch
in von dir vn*d* vrbering reck dich hind*er*sich vn*d*
secz dein rechte hant auf sein tencke pey seine*m*
glenck vn*d* dein tencke ~~an~~ vnder sein rechte als
hie gemalt stet vn*d* faß dein paid hent in einand*er*
so magstu im dÿ arm*m* prech*en* od*er* er müß dich lassen

Item: When you wrestle with another, but he has both arms under and are a much stronger man and lifts you up and will slam or throw you to the earth with violence, or go around with you, or intends to squeeze you hard, as soon as you realize he wants to injure you, then grip him quickly and strongly with both hands to his head and press him a little to the back and press down, as is depicted here, so you jolt him, with the head and the neck into the torso[55] so that he cannot speak.

Item: If another has also assailed you and brought both arms under, then see that you sink solidly down and twist out of him and when he wants to pull you to himself and intends to lift you up, then lock him, controlling him with both of your arms around both of his arms and squeezing tightly, and set both fists into his stomach, as is depicted here, and also press him away from yourself.

And watch if you need to step under or if you need to come into the hip.

Item: Then if another assails you and brings both arms under, then hold yourself firmly to him with both arms over around the neck, and wrap around him with your right foot as well, outside around his left to the inside, and grasp with the one hand which is outside around his head and hold him by the nose, as is depicted here, and with it, pull him backwards, and set one foot in front of him so you throw him to the back.

Item: If another wrestles with you, holding you by the doublet with both hands, thus on the hips, and flings you back and forth and holds you so strongly that you cannot get to him anywhere, so grip him a while to the inside of his elbow and press in from yourself and unexpectedly extend yourself backwards and set your right hand to his left by his wrist and your left under his right, as is depicted here, and clasp both of your hands together so that you break the arm or he must let you go.

[55] HENNING 2014: 43.

55r Item fast dich ain*er* mit sein*er* rechten hant peÿ dem
goldir vnd mit der tencken vn*n*den pey d*er* joppen
vn*d* helt dich starck also vn*d* wil dich werffen so
greif mit dein*er* ~~tencken~~ rechten[56] hant im in sein rechte
in das glenck vn*d* reib im den arm*m* vmb vnd
greif mit der ~~rechten~~ tencke*n*[57] hant an seine*n* elpogen
vn*d* heb in auf vn*d* secz dan deine*n* ~~rechten~~ tencke*n*[58] arm*m*
auf den seine*n* vn*d* stoß nid*er* das get zu p*aiden* seiten
das ist dar vmb also geschriben daß nit ein
ÿeder verten süll wan es ist ein v*er*porgens stuck

55v Item halt dich ain*er* mit paiden henden ~~peÿ den~~ ainew
in dem goldir vn*d* die ander vn*d* in der joppen als
da hinden gemalt stet magstu im dan dy hant im*m*
goldir nit auß prechen so greiff im mit dein*er* hent
von aussen vnden vmb sein hant vn*d* mit der andern
hant ÿnnen vn*d* sleusß paid hent inn anderen als
hie gemalt stet vn*d* heb dich mit dem haupt
hindersich vn*d* heb dan mit paiden arm*m* seinen
arm*m* auf nachot pey den*n* glid so prichstu ÿme*n*
das get auch zw payden seÿtten vn*d* ist güt

56r Item halt dich ain*er* mit paiden henden peÿ denn
achsselen oder peÿ dem goldir vn*d* zeucht dich
starck nid*er* so sleusß dein paid hent in ein ander
vn*d* scheub in ein weil von dir vn*d* zuck in dann
vrbering an dich vn*d* stosß im mit deine*n* paiden
henden vnden auf an sein arm*m* als hie gemalt
stet vn*d* vall im dan nach den fussen oder nach
dem ruck vn*d* wart wie du ein vortail vinden
mügst dastu in gewerffen mügst vn*d* vbereylen

[56] *rechten* added above the deletion.

[57] *tencken* added above the deletion.

[58] *tencken* added above the deletion.

Item: If another holds you by the doublet, with his right hand at the collar and with the left below and thus holds you strongly and wants to throw you, then grip him with your ~~left~~ right hand to his right at the wrist, and grind around his arm and grip with your ~~right~~ left hand onto his elbow and lift him up and set your left arm onto his and strike down.

Do this on both sides.

It is a concealed technique, therefore, it is written that not everyone should know it.[59]

Item: If another holds you with both hands, one onto the collar and the other below to the doublet, as was previously depicted, then if you cannot break his hand free from your collar, then grip him with your hand from the outside, under around his hand, and with the other hand inside, and lock both hands into each other, as is depicted here, and then heave with the head backwards, and lift his arm up with both arms next to the wrist. Thus you break anyone.

Do this on both sides and it is good.

Item: If another holds you with both hands by the shoulders or at the collar and pulls you strongly down, then lock both hands into each other and shove him away from yourself a while, and then pull in abruptly to yourself and hit him with both arms onto his arms from below, as is depicted here and topple him towards the feet or towards the back and watch how you may find an advantage that you may throw and assail [him].

[59] This is a fascinating series of corrections, wherein the scribe began describing the technique with the opponent on the opposite side. Rather than correcting the opponent's arm, which would result in the technique matching the drawing, the scribe changed the thrower's side. As a result, the technique is correct, but described on the opposite side as is drawn.

56v ~~siehz stuck an~~[60]

Item aber ein gucz stuck helt dich ain*er* peÿ dem
goldir mit dem rechten arm*m*[61] vn*d* scheubt dich starck
von im vn*d* ist dir zw starck dastu im nit auß
magst prechen so greif im mit dein*er* tencke*n* hant
an seinen rechten elpogen vn*d* mit deine*r* rechten
hant greif im voren an sein rechte hant an das
glenck peÿ dem goldir als hie gemalt stet vn*d*
zeuck den hind*er*sich zuck[62] ruck dastu im den arm*m*
an reckst vn*d* stoß[63] ÿme*n* in dem glid den enczbaÿ

57r Item so dich ain*er* mit paiden arm*m* hinder wärtling vm*b*
vächt vn*d* dich starck druckt zw im vn*d* hebt dich auf
vn*d* wil dich werffen als pald er dich also auf hept
so greif mit deine*n* paiden henden v̈bersich v̈bersich
auf vn*d* greif hindersich nach seine*m* kopf vnd
vasß in pey dem har als hie gemalt stet vnd
würff in v̈ber dein haupt aus vnd zeuch starck
so müesß er her v̈ber

57v Item so dich nün ainer also peÿ dem har gevast hast
vnd dich her v̈ber ziechen will so greiff im mit
deinen paiden henden oben an sein elpogenn
vnd spann inn hindersich als hie gemalt stet
so magstu in woll also halten wie lang dw
wild oder du magst in hindersich auf denn
rucken werffen als es dan oft zu söliche*m* ku*m*pt

58r Item mer ein stuck wen dich ain*er* also peÿ dem har
wil neme*n* so zuck das haupt hindersich das ers
nit erraichen müg so er dir dan das haupt also
zuckt so puck dich pald vor nider vn*d* vach in mit
deine*n* paiden henden zwischen dein*er* fuß seinen
füß als hie gemalt stet vn*d* heb in starck auff
so würfstu in hindersich auf den ruck

[60] From a different hand.

[61] *armm* added above the line.

[62] Should read *zu*. The scribe stopped before finishing the letter 'k' but did not delete his error.

[63] *stoß* added above the line.

~~look at the technique~~

Item: Once more, a good technique. If another holds you by the collar with the right arm and shoves you strongly from him and too strong so that you cannot break him off, then grab him with your left hand outside his right elbow and grasp him with your right hand holding onto his right hand onto the wrist at the collar, as is depicted here, and pull him backwards behind yourself that you stretch his arm and strike anyone in the limb [breaking it] in half.[64]

Item: Then if another has gripped around you from behind with both arms and presses you strongly to himself, and lifts you up and wants to throw you, then as soon as he lifts up, then grab with both of your hands up over yourself and grasp backwards towards his head and hold him by the hair, as is depicted here, and throw him up over your head and pull strongly so he must go over.

Item: As now another has grasped you by the hair and wants to pull you over, then grab him with both hands up on his elbows and clamp them together behind him, as is depicted here, so you may well hold him as long as you wish, or you may throw him backwards to the back, as this often comes to such.

Item, one more technique: Then when another wants to take you by the hair, then pull your head backwards so that he cannot reach it.[65]

Then if he pulls the head thus to you, then dive down in front quickly and grasp his foot with both hands between your feet , as is depicted here, and lift him up strongly, so you throw him backwards to his back.

---

[64] Note: The technique is described on the opposite side from the drawing, which shows the strong attacker holding with the left hand at the neck.

[65] This is countered by 59r.

58v It*em* so du nün ainenn also hintersich gevast hast
vn*d* er wil dich pey dem har hin über ziechen oder
wil dich peÿ dem fusß ergreiffen als pald er
dich peÿ deinem fusß ergreift so stosß in mit
deinen paÿden hend*en* hinden auf denn ars
als hie gemalt stet so velt er fürsich auf das maul

59r It*em* so dich ainer also hint*en* vmb vächt als vor
geschrib*en* stet vn*d* wil dich zu dem haupt nit
lassen komen vnd wil dich auch zw denn fussen
nit lassen kum*m*en so dauch in ob*en* ein wenig
mit deine*n* schulter*en* hindersich vn*d* in dem selb*en*
so trit[66] mit deine*m* tencken fusß hinder seinen rechte*n*
fuß als hie gemalt stet so würfstu in auf denn
ruck gleich als auf d*er* twirch das get zw p*aiden* seitt*en*

59v It*em* so dich ainer also hinden vmb vangen hat vn*d*
helt dich also vast dastu zw kainen stuck kumen
magst weder zw dem haupt noch zw den fussen
noch in dÿ twirch so vasß in[67] peÿ dem dawm oder peÿ
einem ander*en* vinger als hie gemalt stet so müß
er dich lassen vnd magst im den vinger auß
dem glid prech*en* vn*d* ist ein v*er*porgns stuck

60r It*em* so du ainen also vmb vangen hast so halt in
vast vn*d* secz dich gewiß in dÿ wag vn*d* hütt deins
kops vn*d* deiner fuß vn*d* vor der tbirch vn*d* sleus auch
dein feust über einand*er* das er kaine*n* vinger er-
greiffen mög vn*d* ob er nün ain ving*er* oder dawm
ergriff vn*d* peügt dirn*n* zuruck so gib im pald
nach vn*d* greiff mit deiner ander*en* hant zw vor*en*
in sein glenck vn*d* ker dich vor im vmb vn*d* zeuchs
auf dein achssel als hie gemalt stet so stostu im
den arm*m* ab das get auch von paiden seiten*n*

[66] *trit* added above the line.

[67] *in* added above the line.

Item: Now you have grasped another from behind him, and he intends to pull you over by the hair or wants to seize you by the foot, as soon as he has clasped your foot, then push him with both hands behind onto the buttocks, as is depicted here, so he falls forwards onto the mouth.

Item: Then if another has grasped around you from behind, as written before, and he does not want to let you come to the head, and he also does not want to let you come to their foot, then press him backwards with your shoulders a while, and in the same movement, then step with your left foot behind his right foot, as is depicted here, so you throw him to his back, the same as in the Crossing.[68]

Do this on both sides.

Item: Thus if another has clasped you from behind and then also holds you firmly so that you may not come to a technique, neither to the head nor to the foot nor into the Crossing, then grasp him by the thumb or by one other finger, as is depicted here, so he must let you go, and you may break the finger from the joint.

And this is a hidden technique.

Item: Then again, when you have clasped around another, then hold him firmly and set yourself certainly in the balance, and protect your head, and your foot, and against the Crossing, and also lock your fists over each other so that he may not grasp a finger.

Now, if he grasps a finger or thumb and bends it backwards towards you, then give it quickly towards him and grasp with your other hand forward to his wrist and turn yourself around before him and pull onto your shoulder, as is depicted here, so you break his arm off.

Do this on both sides.

[68] See 17v, the First Crossing.

60v It*em* so es sich also schickt in den ringen oder
in ander*en* ringen das dich ainer also gevast
hat denn arm*m* vber dy achssel als hinden ge-
malt stet wen dw siechst das er sich naigt
nach dem arm*m* so zuck in starck hinder sich zu
ruck als hie gemalt stet oder stosß in mit eine*m*
fuß hinden in den rucken das in vil stuck*en* nücz

61r It*em* so dich ain*er* mit paid*en* arm*en* vor*en* vmb va*n*ge*n*
hat vn*d* helt dich starck so greif mit einer
hant welches ist vor*en* in das goldir ~~vnd~~ als
hie gemalt stet vn*d* secz[69] im eine*n* vinger in der dreier
löchlein ains oder grublin dÿ vor*en* an dem
hals stend vn*d* reck dan den ving*er* vn*d* stich
in da mit in den hals dastüt gar wee

61v It*em* du mit aine*m* ringst vn*d* er vb*er*eilt dich vn*d*
vmb vächt dich also das er dir dein paid arm*m*
pschleust vn*d* zam druckt als oft aine*m* in einer
schied geschicht vn*d* gehalt*en* wirt so dich nün
ainer also vast so senck dich vrbering nider vn*d*
siech ob du in mit eine*m* knie vell*en* mügst so weicht
er mit den füssen auseinander so stoß in dan mit
ein knie in dÿ hod*en* als hie gem*a*lt stet so let er dich

62r It*em* ob sich nün pegäb mit ringen oder in einer
schied das dich ainer hinder wiercz also vmb ving
so sich dastu in ein stoß mit dem kopf gebst in das
angesicht in der selben weil v*er*such ob du in hint*er*
tret*en* mügst als in dem segst*en* stuck da hind*en* stet
stet er dan starck vn*d* helt dich fast so greiff in
mit deine*r* tencken hant peÿ seine*n* h*oden* vn*d* halt
in starck als hÿ gemalt stet so muesß er dich lassen

[69] *secz* added above the line.

Item: If it also arises in that wrestling technique or in other wrestling techniques that another has thus grasped you with the arm over the shoulder as previously depicted, then when you see that he draws in the arm, then pull him strongly backwards to his back, as is depicted here, or strike him with a foot behind in the back.

This is useful in many techniques.

Item: If another has clasped around you with both arms from the front and holds you strongly, then grab his collar with a hand, whichever is forward, as is depicted here, and set a finger into his triple little hole,[70] or the small pit in the front of the throat and then extend the finger and thrust with that into the throat.

That causes great pain.

Item: If you wrestle with another and he has assailed you and clasped around you, so that he presses with his arms locked and controlling you, as often happens in a separation,[71] and you are held. So now, when another holds you that way, then sink yourself abruptly down and see if you may fell him with a knee. If he gives way with the feet apart, then strike in with the knee into the testicles, as is depicted here, so he lets [go of] you.

Item: Now, if it happens in wrestling or in a separation such that he has again engaged, and has clasped you from behind, then see that you give him a strike with your head into the face, in the same moment try, if you may, to step behind him as in the sixth technique.[72]

If he stands strongly behind you and holds you firmly, then grip in with your left hand by his testicles and hold him strongly, as is depicted here, so he must let you [go].

[70] This is a strange phrase, but seems to refer to the sternal notch at the base of the throat. The phrase *oder grublin dÿ voren an dem hals stend* could be seen to indicate another way to describe the same thing as the "triple little hole" rather than as an option opposed to the "triple little hole."

[71] *Schied* is tricky to interpret, meaning a physical or legal separation. It could imply a quarrel, or even a robbery (being separated from one's items) but the true meaning here is obscure, so I went with the neutrally correct meaning.

[72] This reference is confusing, as the technique seems to be describing the stepping behind described in 59r. However, on 59r there is a reference to throwing the opponent "as from the Crossing", which seems to be referencing the very first Crossing (*Zwerchringen*)

62v Item so es sich füeget dastu gevangen wurczt oder
gehalten als das dich ain*er* mit seiner tencken hant
hinten pey dem goldir fassät vn*d* dich starck helt
so ker dich gegen im auf dein rechte seiten vn*d* gee
durch mit deine*m* rechten arm*m* vn*d* greiff mit d*er*
tencke*n* hant in sein tenck hant voren in das glenck
vn*d* mit dem rechten auf in dÿ höch ŭb*er* seine*n* elpogen
vn*d* trit für mit dem rechte*n* fuß als hie gemalt stet

63r Item so dich ain*er* also helt hinden pey dem goldir
als vor so lasß dich ein weil also halten oder füren
vrbering so ker dich gegen im auf dein rechte seiten
als welstu mit im reden vn*d* puck dich mit dem
haupt reschlich vor nider vn*d* stosß in in denn
pauch als hie gemalt stet so velt er an den ruck
helt er dich dan mit d*er* rechte*n* ha*n*t so gecz auf di and*er* seiten

63v Item ein anders stuck wen dich ain*er* hinten peÿ
dem goldir helt so ker dich auf dein tencke
seÿtten von im vn*d* stosß in mit dein*er* rechten hant
an seine*n* tencken elpogen als hie gemalt stet
so müs er dich lassen das get auch zu paid*en seiten*

64r Item mer ein stuck ist dastu mit aine*m* kriegst
vn*d* vor im stest so lege dein hent auf deinen
pauch vn*d* dy recht oben slecht er dir dan zu
mit der fäust gegen dem gesicht so var auf
mit deine*r* rechten hant vn*d* slach im zu sein*er*
feüst mit offenr hent als hie gemalt stet

on 17v. This can be counted as the 7th technique from the front, if one counts the 3 lessons on Wrestling in the forward, or the fourth described technique. The only conclusion I can reasonably come to is that, sadly, our manuscript must be missing either two techniques at the front, or we are missing some other indexing that would make this reference clear.

Item: If it happens that you are captured or held by someone such that his left hand grips you from behind by the collar, and you are held strongly, then turn yourself to him going toward your right side and go through with your right arm and grab his left hand in front towards the wrist with your left hand and with the right hand up high across his elbow and step in front with the right foot, as is depicted here.

Item: If another also holds you from behind by the collar as before, then let yourself be held or driven around a while, then abruptly turn yourself to him toward your right side, as if you wanted to speak to him, and duck down quickly with your head low in front and strike him in the stomach, as is depicted here, so he falls to the back.[73] If he holds you with his right hand, then go to the other side.

Item: Another technique: when another holds you from behind by the collar, then turn yourself to your left side, away from him, and strike in with your right hand onto his left elbow, as is depicted here, so he must let you go.[74]

Do this on both sides.

Item: A further technique for when you fight[75] with another and stand before him, then lay your hand to your stomach and the right over it. If he strikes with the fist to your face, then drive up with your right hand and strike him to his fist with an open hand, as is depicted here.

[73] The image depicts a leg grab, which is not described in the text, but which would make the technique more likely to result in a fall.

[74] This is quite similar to Kal's "Striking Away" (PKM, f 81v), here applied to a Collar Grip rather than to an equal grip.

[75] The manuscript shifts from discussing *ringen* to *kriegen* here and the implication is that the exchange and the resulting context is quite different. I alternately translated this as "fighting" or "brawling" and in either case, it seems that the context is that it is reasonable to strike and kick the opponent in ways that appear to be left out of the earlier material. Whether this indicates a "bar fight", a "self-defense", or a "military skirmishing" situation is not expanded upon by the text and so we are simply left to our own imaginations on the particulars of this context.

64v It*em* mer ein ſtuck so dich ain*er* also in das maul
will slach*en* so var mit der recht*en* recht*en* hant
auf als vor vn*d* trit mit dem recht*en* fuß hinder
seine*n* tencke*n* vnd würf in v̈b*er* das knie als hie
gm*a*lt ſtet so magſtu im auch ein mordſtuck thün

65r It*em* mer ein ſtuck so dich ain*er* also mit der
feüſt slach*en* wil so slach im auch nach seiner
rechten feüſt mit dein*er* tencke*n*[76] von aussen zu vn*d* trit mit deine*m*
tencke*n* füsß hint*er* seine*n* rechte*n* als hie gemalt
ſtet so[77] würfſtu in auch an den ruck als vor

65v It*em* mer ein ſtuck wenn dir ain*er* nach dem ange-
sicht slecht mit der feüſt so vach den slag in dein
rechte hant vn*d* ſtosß in auf den elpogen mit dein*er*
tencke*n* hant als hie gem*a*lt ſtet so p*r*ichſtu im de*n* arm

66r It*em* mer ein ſtuck wen dir ain*er* zw dem ge-
sicht slecht mit der feüſt so var im entgegen
mit dein*er* ~~rech~~ tencken hant vn*d* wint im dein
tencke hant vmb sein rechte*n* arm*m* vn*d* druck
in hint*er* dein tencks v̈gssen als hie gemalt ſtet
vn*d* gib im ein mordſtuck mit der rechte*n* hantt

66v It*em* mer ein ſtuck wen dich ain*er* mit der feüſt
wil slach*en* so var mit deinem gereckten tencken
arm*m* im gegen seine*n* hals mit der feüſt als hÿe
gemalt ſtet vn*d* puck das haupt vaſt nider
vnt*er* den tencke*n* arm*m* vn*d* mit d*er* recht*en* faüſt
ein mordſtuck oder nach seine*m* rechten füsß

67r It*em* mer ein ſtuck wen ain*er* vor dir ſtet vnd
mit dir kriegt so var im mit deine*n* paiden he*n*d*en*
nach seine*m* angesicht so vert er dir auf nach den
hend*en* als well ers wer*en* so ſtosß in mit dem knie
in dÿe ho*den* alz hie gem*a*lt ſtet so müß er an ruck vall*en*

67v It*em* mer ein ſtuck so dir ainer nach dem kopf
oder angesicht slecht so heb deine*n* füsß auff
welchers iſt der dir am peſt*en* fügt vn*d* ſtosß in
in de*n* pauch als hie gemalt ſtet so velt er

[76] *mit deiner tencken* added above the line.
[77] *so* added above the line.

Item: A further technique when another wants to strike you in the mouth, then drive in with your right hand, as before, and step with the right foot behind his left and throw him over the knee, as is depicted here, so you may then also perform a deadly technique[78] to him.

Item: Another technique when another wants to strike with their fist, then also strike him to his right fist with your left to him from the outside, and step with your left foot behind his right, as is depicted here, so you throw him also onto the back, as before.

Item, further technique: When another wants to strike you with his fist to your face, then catch the hit in your right hand and strike in onto the elbow with your left hand, as is depicted here, so you break his arms.

Item, a further technique: When another wants to strike you with their fist to your face, then drive against it with your left hand and wind your left hand around his right arm and press in behind your left armpit, as is depicted here, and give him a deadly technique with the right hand.

Item, a further technique: When another wants to strike you with their fist, then drive with your left arm outstretched, going into his neck with your fist, as is depicted here, and duck your head quite low under your left arm and [drive] with your right fist [either] a deadly technique or to his right foot.

Item, a further technique: When another stands before you and is fighting with you, then drive in with both of your hands to his face, so he will track up after your hands, because he wants to defend, then strike him with your knee into the testicles, as is depicted here, so he must fall onto his back.

Item, a further technique: If another punches at your head or face, then lift your foot up (whichever is the best fit) and strike him in the stomach, as is depicted here, so he falls.

[78] *Mordstuck* can be anything from a kick to a punch to a hit with the hilt of the sword. What this seems to be indicating is that we are in a situation where it is reasonable and appropriate to strike to injure or kill.

68r Item mer ein stuck wen dir ainer nach dem maul
slecht mit der feüst so var auf mit dein*er* tencken
hant vn*d* vach den slag in das glenck vn*d* greif
mit den vingeren vmb sein arm*m* vn*d* greif mit d*er*
recht*e*n ha*n*t vnd*en* nach seine*m* elpogen alz hie gemalt
stet vn*d* zeuch dan so prichstu im den arm*m* ab

68v Item so dir ainer zw slecht mit der feüst so var
auf mit der rechten abichen hant vn*d* vach
den slag inne*n* in dÿ hant vn*d* hab dan vast
zw vn*d* greif mit der tencke*n* an seine*n* elpogen
vn*d* heb yme*n* auf indhöch als hie gemalt stet
vn*d* trit mit dem te*n*cke*n* fuß für in so würfstu
in vber den füß vn*d* prichst im den arm*m* ab

69r Item aber ein stuck wenn dir ain*er* zw slecht so
var aber auf alz vor mit der rechten hant vn*d*
scheub in vast zu ruck auf indhöch vn*d* greiff im
mit dein*er* dein*er* ~~rechten~~ tencke*n* hant hinden
vmb sein rechten arm*m* dürch deine*n* rechten arm*m*
vn*d* trit mit dem rechten fuß hint*er* seine*n* rechte*n*
fuß als hie gemalt stet so prichstu im den arm*m* ab

69v Item mer ein stuck wen dir ain*er* zu dem maul
wil slachen so var auf mit der rechten hant vnd
vach den slag auf denn arm*m* vn*d* stosß in dann
mit dein*er* tencke*n* hant vnden an sein ÿgssen
als hie gemalt stet so stostu in auf den ruck

70r Item mer ein stuck so dir ainer pöse wort gibt
so thü als welstu in mit dein*er* rechten hant an das
or slachen so zuckt er den ~~fuß~~ haupt an weg
mit dem slach in mit deine*m* tencke*n* fuß an seine*n*
rechte*n* füß alz hie gemalt stet so velt er an rucke*n*

70v Item so du mit aine*m* raufst vn*d* er dir vast
zw starck ist so vasß in mit der rechten hant
pey dem har vn*d* mit der tencke*n* stoß in in dÿ
zend alz hie gemalt stet so müs er dich lassen

Item, a further technique: When someone punches at your mouth with their fist, then drive up with your left hand and catch the hit in the wrist, and grasp with your fingers around his arm, and grasp with your right hand under to his elbow, as is depicted here, and then pull so you break his arm off.

Item: If another comes to hit you with a fist, then drive up with the inverted right hand and catch the punch inside the hand and hold it tightly and grasp with your left onto his elbow and lift anyone up upward,[79] as is depicted here, and step with the left foot in front of him, so you throw him over the foot and break his arm off.

Item, another technique: When another punches at you, then drive up again, as before, with your right hand and shove him firmly back upwards, and grasp with your left hand around behind his right arm, through your right arm, and step with the right foot behind his right foot, as is depicted here, so that you break his arm off.

Item, a further technique: When another will hit you in the mouth, then drive up with your right hand and catch the hit on the arm and strike in then with your left hand from below onto his armpit, as depicted here, so you strike him to his back.

Item, a further technique: If another speaks to you evilly, then do as you want and put your right hand on that ear and hit him so he pulls his head away. With that, strike in with your left foot onto his right foot, as depicted here, so he falls onto his back.

Item: If you are brawling with another and he is too strong for you,[80] then catch him with your right hand by the hair and with your left punch him in the teeth, as depicted here, so he must let you go.

[79] I am reading this as *in die Höhe*.

[80] In order to make sense of this and the following few techniques, it appears that the instruction that he is "too strong for you" means he has ahold of your head with both hands and you cannot break free.

71r Item so du mit aine*m* raufst vn*d* des nit kume*n*
magst vn*d* dir zu starck ist so greif mit dein*er*
tencke*n* hant über sein paid arm*m* ~~als hie~~ vnd
greiff im in sein ~~rechte~~ tencke[81] hant in denn tienr vn*d*
reib ims vm*b* vn*d* greif mit der rechten hant
in sein tencke*n* elpogen als hie gemalt stet so
prichstu im den arm*m* ab vn*d* pleibt nit ain har i*n* d*er* ha*n*t

71v Item mer ein stuck wen du mit aine*m* raufst
so vaß in gleich als hinden[82] vor gemalt stet vn*d* leg
im dann sein tencke faust in deine*n* rechte*n*
arm*m* auf die mauß vn*d* slach in dan mit der
tencken hant auff seine*n* hals alz hie gmalt stet

72r Item ob sein not tät dastu müst zwen gevangen
halten vngepunden so faß sy also ny*m* den aine*n*
pey dem rechte*n* arm*m* als in dem nagsten stuck
da hinden gemalt stet vn*d* den anderen pey
dem tencken als sy hie gemalt sten so helcztu
sy an sorg als lang das ma*n* dir zu[83] hilf ku*m*pt

72v Item ein pessers stuck ob dw zwen halten wild
oder füren so zu versten dastu gesellen genüg
hast dÿ sy dirs nötten helfen vn*d* die selben
gesellen ander auch vachen müsten so vasß sÿ
also alz hie gemalt stet so fürstu sÿ an sarg ein
halbe meil vn*d* hieten sÿ holt paid harnasch an

73r Item du solt mercken ein gut stuck vn*d* vnt*er*halten
als wen ein starcker velt vn*d* wie in ain*er* halten
sol so ainer auf der erd ligt auf dem angesicht
so zeuch im seine*n* tencken arm*m* hinde*n* hin auß
vn*d* trit im mit deine*m* tencke*n* füß auf dÿ maus
vn*d* slach im dy faust für dein schinpain als hie
gemalt stet so helcztu in vn*d* er schon harnasch an hiet[84]

[81] *tencke* added above the deletion.
[82] *hinden* added above the line.
[83] *zu* added above the line.
[84] Corrected from *het*—'i' added above.

Item: If you are brawling with another and cannot come to this, and he is too strong for you, then grip with your left hand over both of his arms and grip him by his left hand in the palm[85] and grind it around and grip with your right hand onto his left elbow, as depicted here, so you break his arm off, and not a hair stays in his hand.

Item, a further technique: When you are brawling with another, then grip him the same as previously depicted, and then put his left fist onto your right arm on the bicep,[86] and strike him then with your left hand to his neck, as depicted here.

Item: if you have a need that you must hold two captives unbound,[87] so hold them as follows: both taking the one by the right arm, as in the nearby technique [which was] previously depicted, and the other by the left, as is depicted here, so you hold them, without worry, until someone comes to your aid.[88]

Item: A better technique if you will hold or drive two, then be sure that you have enough companions which can come to help you and the same companions had to hold the others as well, and then hold them thus, as depicted here, so you drive them without worry for half a mile, and you have an advantage if they both have armor[89] on.

Item: You should note another good technique and pinning, as when a stronger one falls and how you should hold them. If one is lying on their face on the earth, then pull their left arm out behind and step in with your left foot to the bicep and strike their fist before your shin bone, as is depicted here, so that you hold him and[90] he has bright harness on.

[85] Dr. FLATSCHER suspects that *tienr* may mean thumb, HENNING 2014: 316 has it as hand or palm.

[86] This technique is another potential completion to the previous one.

[87] The implication is that they must be held "without rope", using the "chicken wing" as described in PD (f 50v), *Wiltu aber dich massen / des vahens liecht von dir lassen / ringens denn fure / gefangen hin ane schnure.* See also PKW, which describes it as *vngepunden.*

[88] Despite the lack of clarity, the drawings are the same as 71v, implying that you are holding these men in the "chicken wing" as well.

[89] *Harnasch* may also indicate "handcuffs", but it is also true that these techniques would certainly work better in armor.

[90] Read: "if" (?).

73v It*em* mer ein stuck ob du aine*n* halt*en* wild
so leg in auf den pauch vn*d* stee grietlisch*en*
v̈ber in vn*d* knie mit dem recht*en* knie auf
in vn*d* ny*m* in mit der rechten hant vor*en* peÿ
schopf vn*d* zeuch in v̈bersich so helcztu in als
hie gemalt stet vn*d* ist ein gut gsell*en* stuck

74r It*em* aber gar ein gut vnt*er*halt*en* als wen ain*er*
gevangen hie so magstu dreÿ man*n* also halten
ob sÿ schon harnasch an hiet*en* vn*d* ist auch ein
grosse pein ob man ainem ger*en* wee wolt thün
alz es hie gemalt stenn dez geschicht oft nott

74v It*em* ob dw einenn paur*en* wild schaczen so vasß
im dÿ haut zusame*n* an seine*m* hals vn*d* stich
im dar durch als hie gemalt stet so maint
er man*n* hab im denn hals abgeschitt*en* vnd schat nit

75r Item ab dw ainem pauern[91]

75v–80r *No text*

80v der stennt Im
Langenschwert
seindt 11 bar

81r–98v *No text*

99r Der Ringen seindt
7 bar stennt

99v–100v *No text*

101r 5. bar stennt Im
Ringen

101v–108r *No text*

[91] Difficult to decipher due to an ink stain that covers the entire line.

Item: A further technique if you want to hold another, then lay him on his stomach and stand straddling over him, and kneel with your right knee onto him and take him by the front of the scalp with your right hand and pull him upwards so you hold him, as is depicted here, and this is a good companion's[92] technique.

Item: Another good pinning, as when one is held captured then you can also hold three men if they have bright harness on, and this is also very painful if you are willing to hurt someone, as is depicted here. This situation is often necessary.

Item: If you want to collect taxes[93] from a farmer, then pinch his skin together on his throat and thrust him through there, as is depicted here, so he thinks you have sliced his throat and no harm is done.

Item: If you [...] a farmer[94]

The stances in the bare long sword are 11.

The wrestling is 7 bare stances.

5 bare stances in the wrestling.

[92] Reading *geselle* is tricky in this position. It could indicate this is a "friendly technique", despite how unfriendly it feels. It is similar to what is shown in depictions of "apprentices" fighting, so perhaps that is what this refers to. As in so much in this treatise, the possible interpretations are many.
[93] The language is ambiguous here, and this could mean "rob" as well as "collect taxes". It seems to indicate a euphemism or a joke, but without more information this must remain unclear.
[94] Copied from the previous page.

108v ich min dich[95]

ia ia liber löff[...]

109r Im Lanngen schwert der ſtennt
Seindt Inn als - - - - - 36 -

Der Ringen Inn als bar ſtent - - 95 -

Im dolchen Inn als bar ſtent - 14 -

Im dusseckhen sent bar ſtent - - - 8 -

Im khampf zu fuß a~~lr~~ alerlaj
ſtennt - - - - - - 51

Mer 7 bar ſtent Im Ringen - - - - 7

Mer 5 bar ſtent Im Lamgen [!]
schwert - - - - - - 5

109v–110v *No text*

CVR P. Öttingen
Wallerſtein'sche
Bibliothek[96]

IBR Nr. *3530*
Jahr: *1989*[97]

[95] Also legible as *ich nim dich*, which has a slightly different connotation.

[96] Library stamp.

[97] In a sticker.

I love[98] you!

Yes! Yes, dear suitor...

In the long sword, the stances are thus 36.

The bare wrestling stances thus 95.

In the dagger, the bare stances are thus 14.

In the *Dussack*, the bare stances are 8.

In dueling on foot, all kinds of stances: 51.

More bare stances in wrestling: 7.

More bare stances in long sword: 5.

[98] Or possibly "take"; either way, the implication of the scene is a proposal of marriage.

mass
Itm also aine an punczt an das swert so slach ain lacke
ein mit dem ort zu dem gesicht vnd wint im mit der
kurczen sneid ein in das angesicht als hie gemalt stet
so magstu im ein reissen mit dem knopff oder mit
gewappent hant in das gesicht stossen

Warp and Weft: The *Bauman Fight Book's* Place in the Tapestry of German *Fechtkunst*

Jessica Finley
Ritterkunst Fechtschule

Christian Henry Tobler
Selohaar Fechtschule

What constitutes a martial arts lineage or tradition? What should be classified as a unique tradition, as opposed to a furtherance of, or divergence from, an existing one? And, equally important, what criteria should we consider in trying to answer those questions? Such matters bedevil scholars of late medieval and early renaissance fight books as they seek to contextualize and categorize their contents.

The *Bauman Fight Book* (BF) presents unique challenges in this regard, but these need not be insurmountable, provided one is willing to view the entire tableau from multiple angles. To that end, this essay will consider the question: what is the relationship of this compendium, particularly its later elements, to the tradition of Johannes Liechtenauer? In doing so, we will rely less on seeking direct lineal connections between treatises and manuscripts, and more on an exploration of how martial philosophies and principles are interwoven through the entire corpus of 15th-century German fight books.

1 Bound Together: The Inherent Complications of Compendia

Many surviving German fight book manuscripts are compilations of multiple martial treatises. A number include other, non-martial material, with some including such wide-ranging subjects as to qualify as commonplace books, or *Hausbücher* ("house books"). Still other manuscripts of the genre are compilations of what were once separate manuscripts. BF falls into this latter category.

BF was at some point, perhaps in the early 16th century, assembled from two or three codices. The second half (Part C) of the current convolution is a textless collection of fighting images, possibly from as early as the 1420s, depicting combat in armour, fighting with large dueling shields, unarmoured long sword fencing, and wrestling (*Ringen*). It is one of our earliest sources for German medieval martial arts, with only the *Walpurgis fight book* (LF),[1] the early 14th-century treatise on sword and buckler fencing, definitively predating it.

Without text to better explain the techniques it depicts, this earlier component's place in fight book history is difficult to establish. While a number of its images are easy to correlate with those found elsewhere, just as many don't fit easily into

[1] Formerly catalogued as MS I.33.

any one part of the puzzle.[2] We have accordingly limited our study to the other, later portions currently bound with it.

The later sections, Parts A and B, bound as the first half of the Bauman's compilation, are likely products of the 1470s. These parts may have been separate manuscripts at one time, or they may have been created together. The two sections each have their own scribe and draftsman, which at first would tend to point to separate creations. However, RAINER WELLE notes that the scribe for the weapons treatise added a corrective annotation on f 56v in the wrestling section.[3] Given this, we may assume that parts A and B are either sections of a single work, or are at least *associated*.[4]

2 The Problem of Liechtenauer

The teachings of Johannes Liechtenauer loom so large in the arc of German medieval martial arts that they invite, perhaps overly critical, comparison between them and other works that don't carry his name, or the mnemonic verse that makes up his treatises on the long sword, mounted combat, and armoured foot combat. But what criteria qualify a treatise as 'Liechtenauerian'?

Certainly, something referencing or glossing Liechtenauer's verses will be readily accepted by most scholars as belonging to the "Liechtenauer Tradition". What if, however, the verse is absent, but a given master's treatise is bundled together with a treatment of Liechtenauer's own work? This occurs in a number of manuscripts, markedly so in the *Peter von Danzig Fechtbuch* (PD), where works by Andre Lignitzer, Martin Huntsfeld, Ott Jud, and Peter von Danzig himself appear along with anonymous glosses of Liechtenauer's three verse treatises. The armoured foot combat treatise by von Danzig is a second, and variant, gloss of Liechtenauer's *Harnischfechten* alongside the anonymous one seen earlier in the manuscript. In contrast, the armoured treaties of Lignitzer and Huntsfeld do not cite the verse, but their corpus of techniques have considerable overlap with those glossed in the other two treatises.

Further cementing the proximity of figures like Lignitzer, Huntsfeld, Ott, and Danzig to Liechtenauer are their inclusion in a roster—the *Gesellschaft Liechtenauers*—appearing in the c. 1475 treatises of Paul Kal,[5] a list of masters associated in some way with the grand master (Fig. 1). Without this connection, one might dismiss these masters as outside of Liechtenauer's tradition; after all, the overlap in terminology is limited and the verse is absent.

Still more problematic are those cases where the verse appears, but almost perfunctorily. Master Hans Talhoffer uses tidbits of the verse to caption the long sword plays in his 1467 fight book (HTM), but the techniques he presents have little

[2] See DIERK HAGEDORN's article in this volume (pp 195–209).

[3] WELLE 2014: 61.

[4] See MICHAEL CHIDESTER's article in this volume (pp 1–11).

[5] PKB, f 2r–4r; PKM, f 2r; PKW, f 9r.

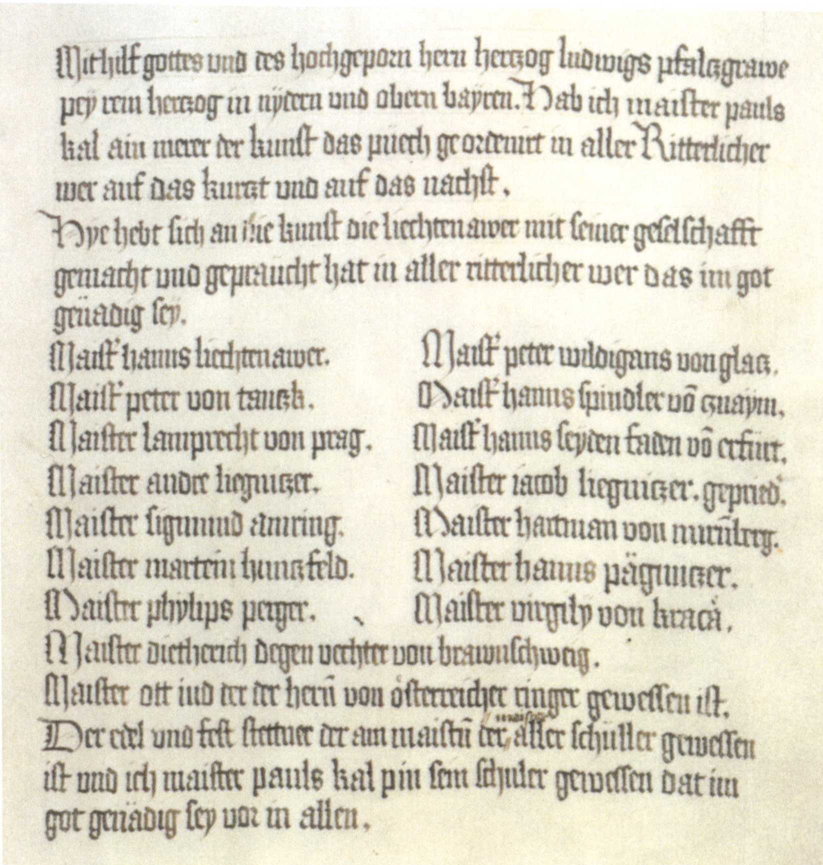
Mit hilf gottes und des hochgepoꝛn herꝛn herczog ludwigs pfalczgrawe pey rein herczog in nydern und obern bayren. Hab ich maister pauls kal ain maister der kunst das puech geoꝛdnirt in aller Ritterlicher wer auf das kurczt und auf das nachst.

Hye hebt sich an die kunst die liechtenawer mit seiner gesellschafft gemacht und gepraucht hat in aller ritterlicher wer das im got genadig sey.

Maist hanns liechtenawer. Maist peter wildigans von glacz.
Maist peter von tanczk. Maist hanns spindler vō cznaym.
Maister lamprecht von prag. Maist hanns seyden faden vō erfurt.
Maister ander liegniczer. Maister iacob liegniczer. gepruedr.
Maister sigmund amring. Maister hartman von nurnberg.
Maister martein hunczfeld. Maister hanns pägnuczer.
Maister phylips perger. Maister virgily von kracā.
Maister dietherich degen vechter von brawnschweig.
Maister ott iud der der herrñ von österreicher ringer gewessen ist.
Der edel und fest stettner der am maisterñ der aller schuller gewessen ist und ich maister pauls kal pin sein schuler gewessen dat im got genadig sey vor in allen.

Figure 1: The *Gesellschaft Liechtenauers* (PKM, f 2r).

direct overlap with those described in the glosses for Liechtenauer's *Blossfechten*. Elsewhere, and conversely, two earlier Talhoffer codices, while containing little or no unarmoured long sword combat, include complete recitations of the *Zettel* for that discipline.[6] The section of PKM addressing the long sword flips this script: there is greater reliance on the verse in captioning his illustrations, but there is no exposition of the entire *Zettel*.

Which of these considerations is paramount? Is citation of the verse, technique overlap, presence in a compendium, or inclusion in Kal's roster the most important qualifier for a work to be considered a part of the tradition? Need we check two of these boxes to convincingly reach such a conclusion? Further, at least for the long sword, there is yet another way to examine the juxtaposition of seemingly outlying

[6] HTG, ff 18r–20v, and HTK, ff 2r–5v.

material—we might also look at such works as *addenda*, or teachings adjunct to that of the main branch of Liechtenauer tradition.

3 Liechtenauer's Terminology in the BF Long Sword Text

Terminology found in Johannes Liechtenauer's verse and its associated glosses does appear in BF, though sparingly. The fundamental principles of *Vor* ("before"), *Nach* ("after"), *Swech* ("weak"), *Sterck* ("strong"), *Indes* ("meanwhile"), and *Fühlen* ("feeling") title folia 4r–6r. For each of these passages, save for the passage on *Fühlen Indes* on f 6r, the accompanying explicatory text also cites these terms.

The passages on *Vor* and *Nach* are of particular interest, as they expand the meaning conveyed in the Ainring, Danzig, and Lew glosses. Those glosses describe acting in the Before or the After only in terms of entering the fight. Bauman's treatment of them extends their meaning by relating the flow of control in the fight to when one has already bound their sword on the opponent's weapon. These plays are therefore a layer built atop material found elsewhere treating the same overarching concept.

Perhaps even more interesting are the first two pages (f 3rv) of the long sword section. These explain the concepts of *Leng* ("length") and *Masß* ("measure"). Liechtenauer concludes his famous prologue with the verse "All arts have Length and Measure", but the main glosses do not explain their meaning. This is notable, for the only other manuscript giving much focus to these concepts is the *Nicolas Pol house book* (N), though in that work Measure (rendered there as *mosse*) receives far more attention than Length.

Other instances of Liechtenauerian terminology include numerous appearances of *Winden* ("winding"), as well as a technique on folio 8r described as an *Ausser Mÿnn* ("outer intention"; Fig. 2):

> Item, another technique. When you strike to his elbow and he then parries the stroke, then remain standing with your sword on his and step forward with the right foot as if you wish to strike to the other side and slice outside to his left ear, as depicted here. This is called the Outer Intention.[7]

This same concept is found in both the verse and in gloss passages such as this, from PD. The specific example described is different, but the conceptual connections are clear:

> **The Technique Described Here is Called the *Aussere Mynn***
>
> *Note:* when he has struck past you and you have chased after him with a stroke to the opening, if he then goes quickly up with his sword and comes under your sword, then remain strongly on it. If he lifts your sword right up with his, then spring with your left foot behind his right and strike him with the *Twerhau* or otherwise to the right side of his head. And work quickly again

[7] BF, f 8r. Translation by Christian Henry Tobler.

Figure 2: The doubling from the outer intention (BF, f 8r).

> against his left side with doubling or else with other techniques, depending on whether you sense he is soft or hard on the sword.[8]

In both cases, the idea of working *outside* the opponent's action is the underlying concept. The *Aussere Mynn* is a specific form of *Nachreisen* ("chasing" or "traveling after"), a way of following the opponent's movement, but performed here in contact with their sword.

4 ELABORATIONS ON THE ART OF THE LONG SWORD

The initial plays in the BF long sword echo and amplify concepts appearing in Liechtenauer's teachings, thereby establishing a connection between the two. In

[8] PD, f 28r. Translation by CHRISTIAN HENRY TOBLER.

Figure 3: A throw with a torso thrust (BF, f 14v).

some cases, they also extend the application of techniques or concepts described in the mainstream glosses. We have already discussed the expansive examples given for the principles of *Vor* and *Nach*. However, it is a passage titled *Masß* which sets the precedent for these expansions. Here, a simple example of *Winden am Schwert* ("winding on the sword") is supplied with follow-on options for closing with the opponent, either in the half-sword grip, here called the *Gewapnoter Hant* ("armed hand"), or by striking in with the pommel. Also noteworthy is the left leg lead shown in the illustration, coupled with a grip with the left hand reversed. These elements extend the basic idea of winding as expressed by Liechtenauer and his commentators.

While in most cases lacking callbacks to Liechtenauer's terminology, the plays that follow the initial seven continue in the same vein of extending an already established art. An entire sequence, stretching from folia 6v to 9v, is a series of counter techniques and contra-counters arising from a simple remedy against the winding on the sword, which is to wind against the opponent's wind.

The focus on answering an opponent's winding action informs a number of the long sword plays. Folio 10v employs a takedown with the half-sword grip following a counter wind, while f 11r moves directly into the half-sword to suppress a winding. These two couple the treatise's concern with countering *Winden* with its considerable exposition of half-sword technique.

The use of the half-sword grip is a hallmark of armoured foot combat in *Fechtbücher* but given little attention in Liechtenauer's *Blossfechten*. Here it appears in eight of the twenty-four plays, perhaps most gruesomely on f 14v (Fig. 3), where it is used to not only throw the opponent by the collar, but to impale their torso while they are on their way down.

BF's focus on unarmoured half-swording is shared by HTM. This is the only Talhoffer manuscript to include substantial long sword combat out of armour.[9] Both codices are examples of what appears to be a growing preference for gripping the blade in the left hand to facilitate close quarters combat, and by including this material each is a furtherance of Liechtenauer's art.

Use of the half-sword grip isn't the only similarity between HTM and the present manuscript. While few of their techniques overlap,[10] each has a limited number of terms drawn from Liechtenauer's *Zettel*. The remainder are unique innovations or variations on actions described in the glosses; many of these techniques focus on grappling with sword in hand. The two are also kindred in that neither presents a complete system of fencing on its own; too much is missing to allow reconstruction of a practicable art of combat. Many a HEMA practitioner decades ago learned the hard way of the perils of trying to learn medieval long sword combat with only HTM to hand; the same holds true for BF. These works are aimed at an audience with a considerable grounding in the art found elsewhere.

5 Dagger and Messer

BF's dagger and Messer treatises neither confirm nor deny connections between it and the mainstream Liechtenauer tradition. This is because, until the advent of Lecküchner's *Messerfechten* treatise, no work on that weapon contained more than a mere handful of techniques, with no terminology overlapping with Liechtenauer's. The same holds true for the dagger, save that no 'magnum opus' was ever created for this weapon.

[9] At least the techniques prior to f 35v appear to be intended for unarmoured fighting, as they include thrusts to areas normally protected by armour such as the chest, abdomen, and forearm. The illustrations following those featuring armoured figures (ff 35v–38r) may be armoured techniques rendered in civilian attire for simplicity's sake. This same pattern appears in HTK, where the armour ceases to be illustrated following a handful of pages where it is depicted. For more on this, see TOBLER 2022: 13–15.

[10] The visual similarities between BF, f 8r, and HTM, f 5v, are compelling.

Figure 4: A feint and strike (BF, f 26r).

Early instruction for Messer and dagger instead takes the form of usually short treatises, some by *Gesellschafts Liechtenauer* authors and some anonymous, that accrete around the main teachings. Dagger works by Lignitzer and Huntsfeld are included in various compendia alongside the glosses and other material deriving immediately from them. They are 'Liechtenauerian' in the sense that they are codicologically associated with other writings more clearly a part of the tradition, and because common principles and sensibilities stand out in them.

The fourteen dagger plays in Bauman are much like those of Peter Falkner, Paul Kal, Martin Huntsfeld, and Andre Lignitzer, as well as the anonymous treatise in the 1508 *Glasgow fight book* (GF). Each of these works has a core of actions common throughout, and handfuls of outliers making them unique. For instance, the feint and groin stab on folio 26r of Bauman (Fig. 4) is unique among 15th-century fight books. Similarly, the few techniques for the "Middle Thrust" are found only in GF's dagger treatise,[11] and slicing cuts to the attacker's hand are described only

[11] GF, ff 99v–100r.

Figure 5: A throw from the armed hand (BF, f 32v).

by Paul Kal.[12] Each treatise has its own unique plays, with nothing so foreign to the core elements as to alienate it from the main tradition.[13]

The situation is much the same for the Messer, only there is very little to work with prior to Leckűchner's 1478 manuscript.[14] Before that, the Messer appears sparsely, finding limited focus in the HTK and HTM as well as comprising a small section of Paul Kal's fight books. The BF Messer treatise is yet another small contribution to this limited corpus, describing only eight techniques, albeit relatively well organized around the employment of three different parries: one with the sword hand turned and the point facing up, the second a "hanging parry", with the third employing the "armed hand", hearkening back to the half-sword techniques of the long sword material.[15]

[12] PKM, ff 75v–76r.

[13] For BF's dagger pieces, see BARTŁOMIEJ WALCZAK's article in this volume (pp 103–127).

[14] *Hans Leckűchner-Heidelberg* (LH)

[15] For BF's Messer pieces, see LUCIEN LEE's article in this volume (pp 129–155).

If the Messer material is difficult to pin to the earlier tradition of Liechtenauer, the *Ringen* section of Bauman's treatise is quite the opposite, as we shall see.

6 CUT FROM THE SAME CLOTH: THE *RINGEN* OF BAUMAN AND OTT

The *Ringen* section, Part B, of the Bauman fight book is easily the most comprehensive and beautifully organized of the currently-studied medieval wrestling treatises. It has the holy grail combination of qualities for a fight book: both an extensive text section and accompanying illustrations, as well as an organizational structure that clarifies techniques as well as providing tactical information for the situation where the technique is most useful. The *Ringen* section opens by explicating a theoretical framework in three lessons, which bear remarkable likeness to that found in Ott's treatise, itself based upon the fight tactics outlined within Liechtenauer's verse.

Liechtenauer instructs us to think on his core concepts, which are Weak and Strong, Before and After, and a fifth which is *Fühlen/Indes* ("feeling" and, loosely, "within"). Weak and Strong are said to be qualities of the two halves of the sword blade, whereas a related idea, Hard and Soft, appear to be qualities of the fencer themselves. Because weaponry is of no consideration in the *Ringen* treatises, both Ott and Bauman use Strong and Weak to apply to the wrestlers themselves, though later wrestling treatises, including Auerswald, apply Strong and Weak to areas of the body as though the body itself is a weapon.

Ott's treatise opens with his lesson on the qualities of wrestling, which he labels as *Kunst* ("skill"), *Schnelligchait* ("quickness"), and *Sterck* ("strength").[16] He suggests that of, these qualities, Quickness is the best as it can prevent you from being countered. Bauman's first lesson labels the three qualities as *Sterck* ("strength"), *Masß* ("measure"), and *Phentikait* ("agility").[17] While BF doesn't declare any of these as best, its anonymous master does tell us a little bit more about what he means by them and how a wrestler might use them. Strength is found by going low in the *Wag* ("balance"), Measure by correctly arranging your hands and feet in the positions you need to achieve, and Agility will allow you to counter and prevent any attacks that come your way. Curiously, we see then that *Masß* can be seen as a synonym of *Kunst*—which is not intuitive, but if we consider *Masß* as a quality of moderation or "measured action" it seems clearer.

Ott's next statement is regarding the question of relative timing of the wrestler's actions: *Vor* ("before"), *Mit* ("simultaneously"), and *Nach* ("after").[18] He ties these times to the question of the relative strength, or perhaps size, of the opponents, thereby giving advice as to how to approach them based upon this assessment: if they are weaker than you, act Before; if they are equal to you, then act Simultaneously; and if they are stronger, then act After. He goes on to say that you should

[16] PD, f 100v.
[17] BF, f 16r.
[18] PD, f 100v.

pay attention to Quickness when you wrestle Before, the Balance when you wrestle Simultaneously, and to the crook of the knee when you wrestle After.

The second lesson in BF deals with the same timing but is a little more specific about what we should be doing in each particular time. If they are weaker than you, then you should wrestle Before using Strength. If they are equal to you, then you should wrestle Simultaneously using Measure. Finally, if they are stronger than you, you should wrestle After using Agility.

He again goes on further with more details than Ott gives us regarding each of these timings and our intention within them. Wrestling someone who is weaker, we should set ourselves well in the Balance, and attack so that he can do nothing but escaping and dodging. Wrestling someone who is our equal, we should also plan to defend ourselves, and to try to get inside his defenses.

Curiously, the instruction on strength is labeled the third lesson, but it clearly continues from these bits of advice. Wrestling someone who is stronger than us, we should beware when they grab us and be ready to break free, if possible. And we should yield and make distance as much as possible so as to not be grabbed, try to mislead them, and attack them by grabbing the foot (or leg).

Finally, Bauman gives us a foreword to the techniques, which is a lesson about the difference between wrestling for fun, and for a deadly situation, which is important enough, and particularly relevant to modern sportspeople, to quote in full:

> Though each weak opponent who wrestles a strong one in earnest is equally likely, if they have the Agility and Measure, fighting techniques and killing techniques, to take the advantage; in friendly wrestling, however, the strong always have an advantage. Nevertheless, the art is praised as worthy by knights and squires[19] beyond all things, etc.[20]

Following this extensive prologue come the individual wrestling techniques, and while these aren't explicitly broken into chapters by the creators of the manuscript, thematic groupings are evident based around the tactical setup. This tactical setup may be related to the timings of Before, Simultaneously, and After, but this thesis needs more research to be confirmed.

The initial situation we are presented with is the entry from extended arms, and the first set of techniques are seven *Twirch Ringen* (or "crossings") with their associated counters and contra-counters, which are laid out in an interconnected framework. With the exception of the first Crossing, each is presented on two facing pages: the initial *Twirch*, a counter technique, and a contra-counter. The first one is exceptional and an exemplar of a deeper wrestling concept—indeed, of a fencing concept—to attack the first available opening and then work with what further openings present themselves as the opponent attempts their counters or denials of the technique (Fig. 6).

[19] "Knechten" is a tricky word to translate without more context as it has flexible meanings such as servant, squire, man-at-arms, hired man, or armed servant.

[20] BF, f 15v. Translation by JESSICA FINLEY.

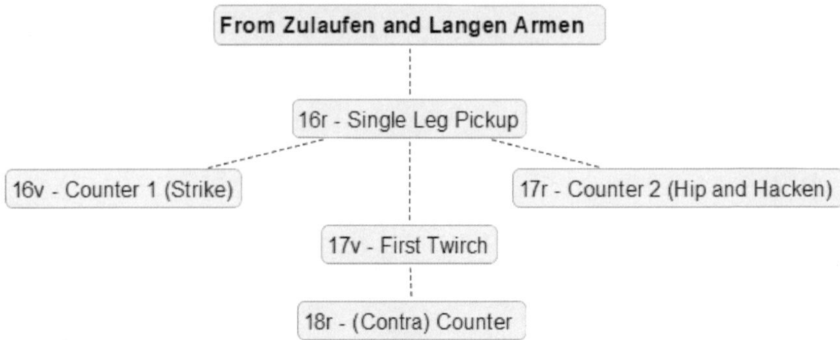

Figure 6: An example of the structure of BF's wrestling.

While the *Twirch Ringen* are relatively unique in their nomenclature and tactical presentation amongst wrestling treatises, not all of the techniques themselves are unique. In fact, the counter to the first *Twirch* is highly reminiscent of a counter found in Ott's treatise. Ott describes the following:

> Note that this technique counters all wrestling from the first stance... and simultaneously send your right arm over his left at his right side, seizing him by the waist. Spring with your right foot behind his left, and throw him off his feet over your right leg.[21]

Bauman's counter reads, in part:

> ...and quickly extend the foot behind him, and with the arm that is against him, grasp up over his shoulder securely forward toward the earth...[22]

After the *Twirch Ringen*, a number of further techniques arise from the *Zulaufen* and *Langen Armen*, and these are extremely common in German wrestling treatises. One example is the *Durchlaufen* ("running through") under the arm, which is the most commonly described wrestling technique. It is similar to a fireman's carry, and there are many variations on this, from performing it while only holding one arm in the entry, to breaking free of a grip with one arm to attack the opening, to doing it at the sword in *Blossfechten*. There are also a series of throws commonly called *Durchwenden* ("turning through"). These appear in Ott, Lignitzer, and Talhoffer amongst others. Again, BF is exceptional not in its techniques, but in the clarity of organization, outlining for the reader a technique, counter, and contra-counter.

[21] PD, f 101v–102r. Translation by Christian Henry Tobler.

[22] BF, f 18r. Translation by Jessica Finley.

Figure 7: A technique from an equal grip (BF, f 44v).

The next section deals with techniques to use from *Gleichen Fassen*, an "equal grip". This is referred to in Ott and other treatises as 'one arm inside and one arm outside', though Lignitzer also offers the description of *Gleichen Fassen* as 'one arm outside and the other under', which makes it clear that the salient point is the reciprocity of the grip, with neither opponent given initial advantage in position, rather than the specifics of what is being grasped. Here again we see many common techniques, with a noteworthy one on f 44v. This throw reads, in BF:

> ...then drive him with your left hand behind the jaw, and with your right between his thighs and grip him by his right thigh and lift up, and then press him over backwards from yourself.[23]

[23] BF, f 44v. Translation by JESSICA FINLEY.

Ott describes it similarly:

> ...send the arm that is above under his chin to his neck, and push him hard away from you. And with the hand that is below, grasp his leg at the crook of the knee and pull with it toward you; thus, he will be thrown.[24]

This common throw appears with variations in dagger, sword, and armored fighting treatises, and is also described in the mini-wrestling treatise *Drei Ringen* ("three wrestlings") which is found in the *Sigmund Ainring fight book* (SR).

Hacken is treated in Talhoffer, but otherwise unknown in the traditionally accepted corpus of Liechtenauerian wrestling, though it does show up extensively in Fabian von Auerswald's later treatise. Whether this is a regional preference or whether it is typically not included in the earlier wrestling material due to its limited usefulness in armor is something that needs further exploration to answer. Still, f 100v of HTM shows a *Hacken* from the front (Fig. 8), which is countered by a "break with the strike". BF's *Hacken* counter on f 48v (Fig. 9) also describes this technique, but in much more detail:

> ...then lift the same foot up and step into his free foot, as is depicted here, and duck down forwards and push him from yourself, and pull his head from your armpit.[25]

The subsequent set of techniques deal with unequal grips at the body, or your doublet and collar being grasped. This setup is dealt with in Ott explicitly and many of the same techniques are employed. For instance, BF describes a counter against a stronger opponent thus:

> ...when he wants to pull you to himself and intends to lift you up, then lock him, controlling him with both of your arms around both of his arms and squeezing tightly, and set both fists into his stomach, as is depicted here, and also press him away from yourself.[26]

Ott says:

> ...go down from above outside and under his elbows, and lift upwards with strength and break his arms.[27]

Bauman's manuscript does conclude with a relatively unique group of techniques dealing with fist strikes. That said, striking isn't entirely unknown in the *Fechtbuch* tradition. In BF, it suggests that the situation we should apply this to has changed, as you are engaged in *Kriegen* (fighting) rather than *Ringen*. Perhaps this is a callback to that foreword stating that there is a difference between the advantages one can apply in earnest fighting versus sportive engagement. In an anonymous

[24] PD, f 103v–104r. Translation by Christian Henry Tobler.

[25] BF, f 48v. Translation by Jessica Finley.

[26] BF, f 53v. Translation by Jessica Finley.

[27] PD, f 106r. Translation by Christian Henry Tobler.

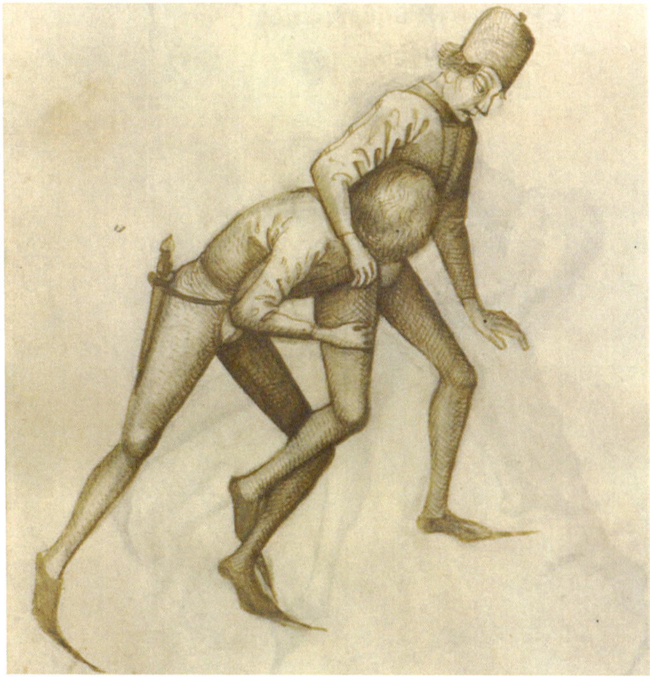

Figure 8: Talhoffer's *Hacken* (HTM, f 100v).

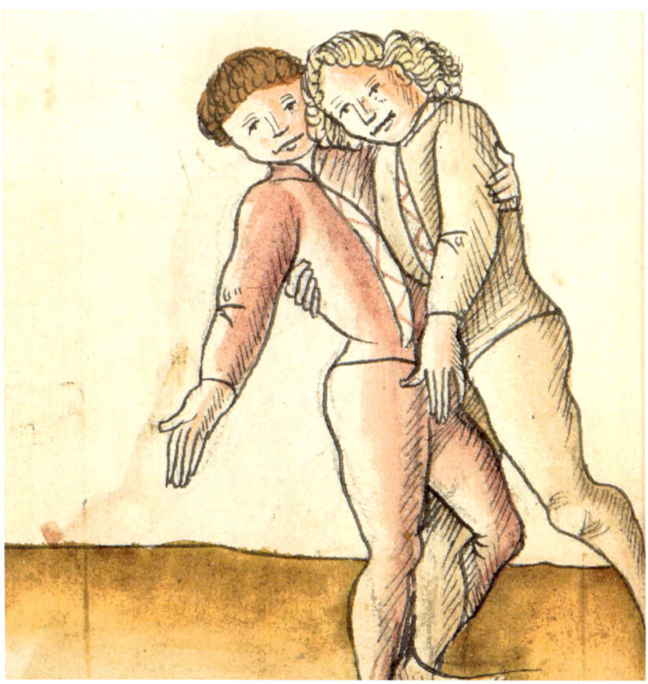

Figure 9: A counter to the *Hacken* (BF, f 48v).

treatise in SR, these strikes are called *mortstossen*, which is a general term applied to both these punching attacks, as well as blows delivered with the hilt of the sword in armored combat. The main difference between the treatise in SF and BF is that BF is almost entirely defending against strikes, while the anonymous treatise gives instruction on how to attack. Even so, we still find overlap in this as the anonymous treatise suggests "Strike him with both fists against the neck as hard as you can. Afterwards wrestle."[28] Bauman instructs that you feint this blow and afterwards knock him down: "...then drive in with both of your hands to his face, so he will track up after your hands, because he wants to defend, then strike him with your knee into the testicles, as is depicted here, so he must fall onto his back."[29]

Despite the fact that unarmed striking appears so rarely, Bauman's tactical framework for dealing with it is utterly in alignment with the dagger treatises that appear throughout the Liechtenauer tradition. First, one must make a cover, whether that be to the inside, outside, or coming underneath the blow, followed by one's own delivery of a blow or wrestling technique to subdue the opponent. An example comes to us from Lignitzer's dagger treatise,

> If he thrusts down from above to your face or breast, then go up from below with your left arm and catch the thrust on your arm. And grasp with your left hand from inside out over his right arm and press it fast to your left side. And then thrust with your dagger to his face.[30]

Bauman gives the same technique against a punch:

> When another wants to strike you with their fist to your face, then drive against it with your left hand and wind your left hand around his right arm and press in behind your left armpit, as is depicted here, and give him a deadly technique with the right hand.[31]

Finally, there are a set holds that appear to be for the purpose of taking prisoners. *Unterhalten* ("holding down") appears in the treatises of the Liechtenauer tradition by authors such as Huntsfeld, Kal, and Falkner, but those plays are related to holding down within a duel between two men and for the purpose of being able to deliver the final blow or receive a concession from the opponent. In this treatise, however, it appears to be more important to be able to restrain multiple people, an idea which also shows up in Talhoffer.[32] Bauman does include a technique on f 73v for holding one opponent, which reads,

> If you want to hold another, then lay him on his stomach and stand straddling over him, and kneel with your right knee onto him and take him by the front

[28] SR, f 76r. Translation by CHRISTIAN HENRY TOBLER.

[29] BF, f 67r. Translation by JESSICA FINLEY.

[30] PD, f 85r. Translation by CHRISTIAN HENRY TOBLER.

[31] BF, f 66r. Translation by JESSICA FINLEY.

[32] For example, HTK, f 138r; compare BF, f 72v.

> of the scalp with your right hand and pull him upwards so you hold him, as is depicted here, and this is a good companion's technique.[33]

It is quite similar to that found in Huntsfeld, though of course he presumes that armor is being worn,

> ...then fall with your right knee behind in his abdomen, and grasp with your left hand over his head in front on the visor, and pull upwards and press down with your knee.[34]

While the *Ringen* section of BF clearly has many unique features, the likeness of the introduction to the opening of Ott's treatise is inescapably telling. Either the author of the *Ringen* section was inspired by Ott's work—or both were working from an unknown original source. In either case, it appears evident that the *Ringen* section is in alignment with the tactics, techniques, and core framework of Liechtenauer's art.

7 Conclusions

We have examined above some of the many connections tying the first, and later-dated, half of the *Bauman fight book* to elements found in manuscripts more commonly associated with the Liechtenauer tradition. These correspondences are found across a spectrum ranging from the conceptual to those where technique directly overlaps.

The Bauman material for the long sword makes direct use of some of Liechtenauer's own terminology before layering on adjunct material, elaborating in a manner not dissimilar from that seen in the work of another master in the Liechtenauer tradition, Hans Talhoffer. The Messer and dagger treatises, while offering some unique innovations, are typical for the short works on those weapons found elsewhere in the 15th century.

The *Ringen* found in BF offers even stronger evidence of direct ties to the mainstream 15th-century fight books, while still offering unique material added to what is found elsewhere. For the former, the correlations between its prologue and that attributed to the wrestling master Ott are profound; for the latter, the techniques for restraining prisoners are a fascinating view into how the overall art applied to the practical world of war and skirmish.

Understanding the place of this work is important, not only from a scholarly vantage, but from a practical one as well. The historical martial arts practitioner is best served by exploring the *Bauman fight book* not in isolation, assuming it is an unrelated outlier, but in its proper context within the arc of a much larger tradition.

---

[33] BF, f 73v. Translation by JESSICA FINLEY.

[34] PD, f 91v. Translation by CHRISTIAN HENRY TOBLER.

8

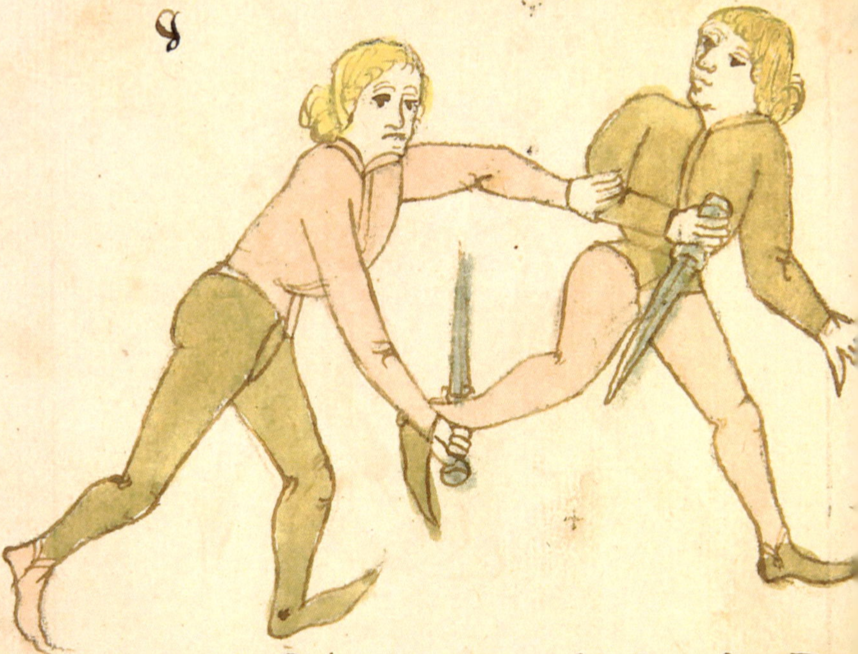

Item sticht dir ainer oben zu dem gesicht so wirff deinen
degen auf deinen rechten arm vnd vach den stich dar ein
vnd stoß im mit der tencken hant an seinen degen vnd
gee dir das erstich war dir vnder vnd fass in bey einem
fuss mit deinem degen alz hie gemalt stet so wirffst

Bauman Dagger Teachings and the Augsburg Group

Bartłomiej Walczak
Unaffiliated

Bauman's Fight Book (BF)—as currently bound—consists of three separate parts, of which two feature elements of dagger combat. Chronologically earlier part C (ff 76r–108r) deals with dagger in the context of armoured judicial duel. The later part A (ff 22r–28v) contains fourteen illustrated dagger techniques that belong to a wider tradition known informally as Augsburg Group or Augsburg Tradition.

Several publications attempted to establish the exact relationship of BF to other manuscripts containing similar teachings. Notably, Welle[1] described an association with the treatise of *Anton Rast* (AR) and Paul Hektor Mair's manuscripts, and later in his critical edition of *Goliath* (G)[2] outlined the connection between it and BF, but could not establish either as an authority. And while the connection between wrestling, long sword, and Messer teachings of BF and similar content found in a manuscript ascribed to *Albrecht Dürer* (AD) has been pointed out in numerous works,[3] the extremely brief dagger section of AD does not seem to share it, with the exception of a single picture in which legs are switched. Finally, *Vechtboek* (V) is a short Dutch manuscript containing several dagger techniques which bears some resemblance to the later part of BF as far as illustration workshop is concerned, but its connection to the Augsburg Group remains tenuous.[4]

In 2017, a more comprehensive list of manuscripts containing teachings that belong to the Augsburg Group was published as part of a global survey of German-language dagger sources.[5]

1 German Dagger Teachings in the 14th–16th Century

Before we talk about dagger teachings contained in BF we need to make several general remarks about how most of German-language sources approach the subject of combat with this weapon.

In general, these teachings are recorded in the following ways:

- Text only
- Picture with detailed textual description
- Picture with a short caption
- Picture only

[1] Welle 2014.
[2] Welle 2017.
[3] Dörnhöffer 1907/1909, Welle 2014, and recently Hagedorn 2021.
[4] See Reinier van Noort and Oskar ter Mors' article in this volume (pp 211–229).
[5] Walczak & Starko 2017.

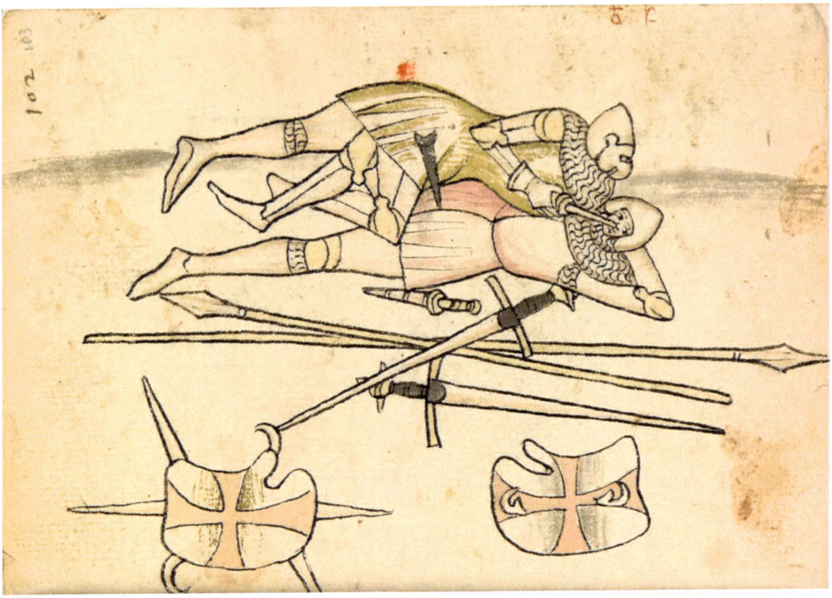

Figure 1: A wrestling play from part C, with weapons scattered on the ground (BF f 103r).

Pictures usually present the key moment of a given technique which would allow the reader to easily identify and recall a technique that they were previously taught. As such, they don't need to be fully anatomically correct, it's enough that they trigger a memory. The downside of this type of recording is that without prior knowledge or detailed description, it can be hard or even impossible to be sure exactly what steps are involved in the execution.

The quality of a picture can therefore be judged not by its aesthetics, but by the uniqueness of the key moment and its ability to easily identify a given technique. For example, Talhoffer's earliest known manuscript contains many pictures of one fighter disarming another that look extremely similar to one another.[6] Of course, we can't know if the person to whom the manuscript was dedicated was able to correctly identify these techniques, but we can see that Talhoffer's later works are much more distinct and therefore arguably better.

Short captions can help a lot with identification and partially with recall of some more complex steps involved in the execution, but require prior knowledge as well to fully recreate presented material. On the other hand, longer textual descriptions give us the most detailed step-by-step execution which is usually pretty straightforward and easy to follow.

Dagger combat techniques are presented as recipes intended to solve or alleviate a particular problem—which is typical for most pragmatic literature of that period.

[6] HTG, ff 83r–96v.

In contrast to, for example, sword material or dagger teachings of *Fiore de'i Liberi* (FLG),[7] the recipes are specific and extremely rarely contain any generalized or abstracted thought. It doesn't mean that the advice cannot be applied elsewhere or that there are no underlying principles governing those proposed solutions, but the authors either did not give it more thought or did not bother to record it. Brief tactical advice can be found in the *Nicolas Pol House Book* (N) as an elaboration of concepts found and explained in its long sword section,[8] but only once does Hans Talhoffer possibly attempt to codify basic attacks used for dagger combat,[9] and even there, commonly used basic defenses are never extracted as separate elements as is the case with other weapons.

With that in mind, let's see how dagger teachings presented in BF fit inside that picture. We will begin the examination from the chronologically older part of the treatise.

2 Overview of Teachings in the *Bauman Fight Book*

2.1 Part C

There is very little that we can say for certain about dagger teachings in part C. Combatants are wearing very late 14th or early 15th century full harness, often with open bascinet helmets or *Hundsgugels*, occasionally having open visors as well. This is similar to what we can see in other sources from this period, such as FLG.[10] Later sources either depict visors fully closed in case of armets and similar helmets (*Gladiatoria*)[11] or open sallets featuring bevors protecting the lower half of combatant's face (Talhoffer, Kal).[12]

The combination of arms (spear, sword, dagger) and riding shields with the cross of St. George, plus the occasional *wappenrock* with the same cross, clearly suggest the context of ritualized judicial combat (Fig. 1). This is corroborated by other plates showing combatants without armor using long shields and wearing special costumes also used in such duels.

The depicted techniques do not seem to be organized in any obvious manner. Spear, sword and dagger are interspersed, though on plates with two techniques different weapons never share the same plate. Considering multiple plates showing dagger use with or against sword[13] or spear against sword,[14] we may also assume that the set of weapons was thought of as a continuum to be employed

[7] FLG, ff 9v–22r.

[8] N, ff 84r–85r.

[9] HTG, ff 83rv.

[10] FLG, ff 32v–35r.

[11] GK, GN, GW (whole manuscripts).

[12] HTK, f 86rv; PKM, f 26r.

[13] BF, ff 81v, 89v, 103v.

[14] BF, f 105r.

whenever situation warranted their use. At the same time, since it is one of the earliest known surviving books of this type, we may speculate that perhaps this was an early experiment produced before more standardized methods of recording and organization were developed.

While several techniques of sword in armor can easily be recognized in another set of treatises named the *Gladiatoria* Group,[15] it is not so with dagger. The material shown here is difficult to find in other sources.

The only exception seems to be the works of Paul Hektor Mair, who in the 16th century obtained the whole codex and, when creating his all-encompassing compendium of knightly arts, adapted eight out of the thirteen techniques, omitting those that feature dagger against sword. His treatise includes very verbose descriptions; however, as we shall see later, they may have very little relevance to the original technique depicted by its author.

Because the number of techniques is small, we can go over them one by one and trace possible references in the later sources. Techniques marked as signatory are uncommon or complex and can serve as the markers of a given tradition. Unique techniques are to my best knowledge present only in this manuscript and perhaps later copied by Mair.

81v A dagger thrust into the face after immobilization with a sword, a signature technique later present in *Gladiatoria* (Fig. 2).[16]

83v A disarm/immobilization using the scissors defence, very common in other sources.

86r Possibly an elbow pull described in *Gladiatoria.*[17]

86v Possibly a lower shield counter by pulling the opponent's head or a preparation for a throw. Unique.

87r Possibly a disarm or a counter to disarm; possibly signature use of the left hand, unique.

87v This technique bears some resemblance to Talhoffer's counters to upper shield,[18] but is unique.

88r-A While the elbow leverage is rather common, lifting the knee for a throw seems to only be present in Talhoffer's final work.[19]

88r-B A throw by pushing the opponent's dagger arm between his legs. The general idea is common, both in wrestling and in dagger, though this particular execution seems unique.

89r Preventing the opponent from drawing the dagger by securing his hand and possibly striking him in the face. Signature, unique.

91v A thrust into the left hand against a half-sword thrust. Signature, unique.

---

[15] See DIERK HAGEDORN's article in this volume (pp 195–209).

[16] GK, f 28v; GN, 22v; GW, 26v.

[17] GK, f 43v; GW, 42r.

[18] HTM, f 93r.

[19] HTM, f 95v.

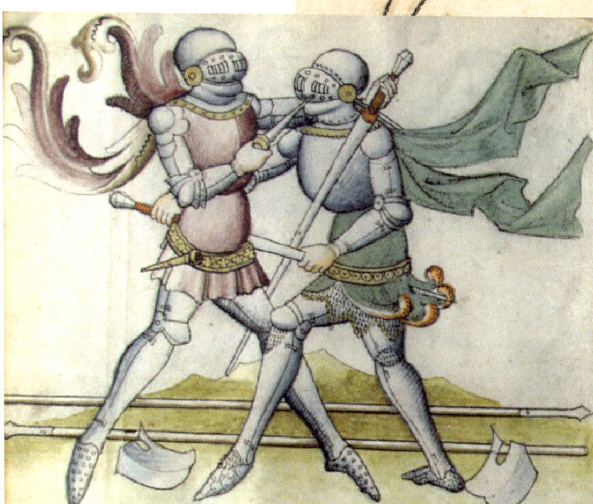

Figure 2: A dagger thrust to the face (top: BF, f 81v; middle: GN, f 22v; bottom: PKD II, f 231v).

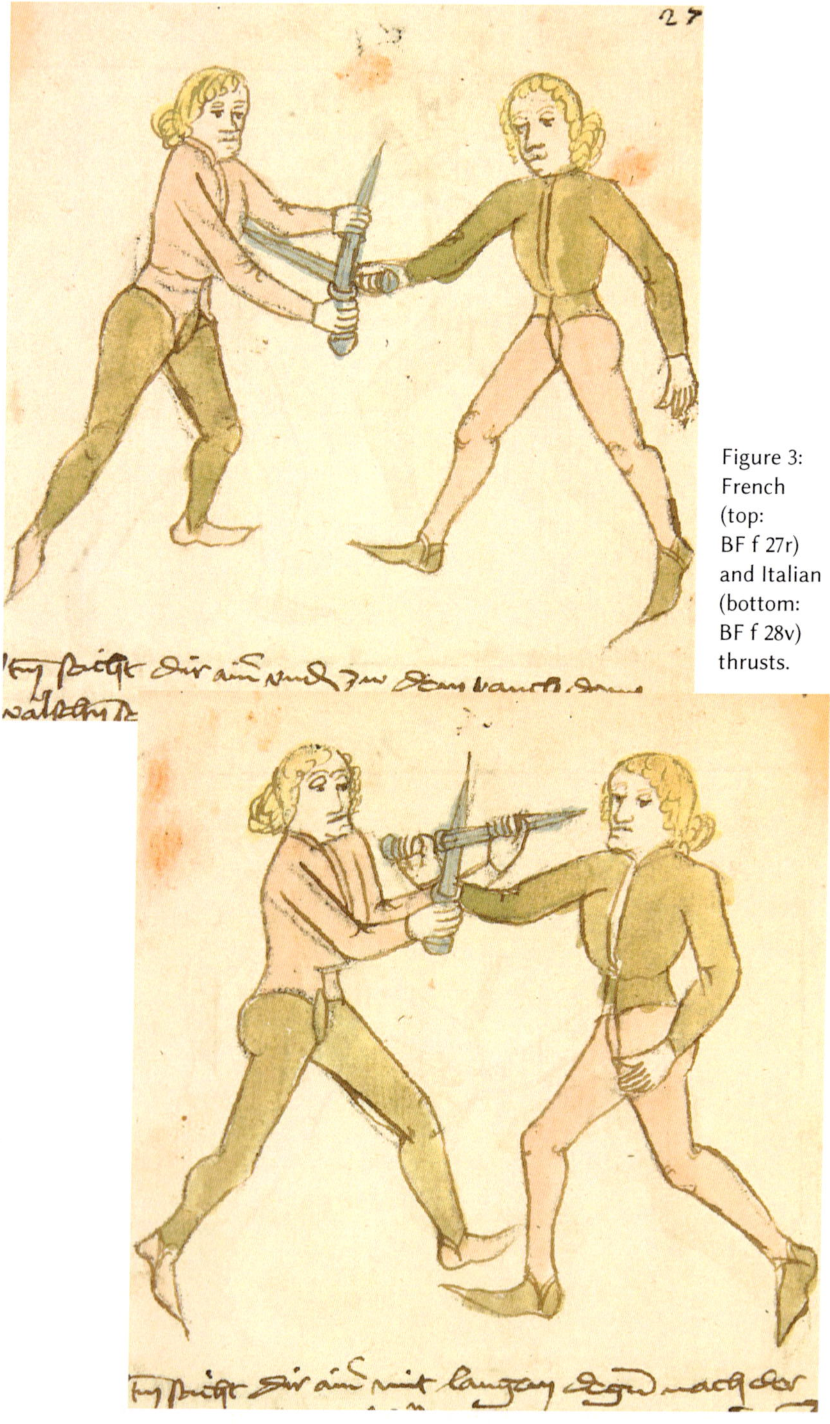

Figure 3: French (top: BF f 27r) and Italian (bottom: BF f 28v) thrusts.

95r A possible disarm or counter, but the interpretation is uncertain. Signature, unique.
103r A thrust to the face after pinning the opponent to the ground (Fig. 1). Seems unique, but simple enough and can't really be marked Signature.
103v Neck scissors against the half-sword. Signature, unique.

Overall, part C of the codex shows an interesting, mostly unique dagger use in the judicial duel.

2.2 PART A

At the first glance, Augsburg teachings included in parts A and B do not seem to form a cohesive whole. It feels like one could easily separate the sword, Messer, and dagger material and it wouldn't make any difference. Especially when it comes to dagger, each technique is different and self-contained, which is pretty typical for the existing material from that period (1470s) and area.

There are fourteen dagger techniques in the manuscript. Presentation follows "picture with description" format and directions are quite straightforward with few exceptions. Combatants are unarmored, but they are using long rondel daggers, typically associated with armored combat. The techniques presented strongly suggest a lack of armor, vulnerability, and unhindered movement. Thrusts are aimed at the face (not the visor), neck, or stomach, while the disarms and locks would be difficult if not impossible when executed in harness.

Techniques are usually quite simple with the exception of the very first one,[20] which contains a feint that probably expects the opponent's reaction since the actual lock is applied in a place which should circumvent the anticipated defense. An oddity is an actual recipe on how to execute an attack with a dagger.[21] With few exceptions, all dagger material is devoted to defense against being attacked, and this one piece in BF shows how to make a non-trivial feinted attack.

It's difficult to judge the criteria of how the techniques were chosen to be included and in what order. There seems to be a lot of arbitrariness in deciding that complex techniques containing reactions to the perceived opponent's initial responses should be put alongside or even before some other more basic techniques, such as a knee throw or scissors defense. Like most of the dagger material from this period it seems opportunistic and unsystematic.

Another interesting element seems to be that two of the thrusts are named after nationalities (Fig. 3): French (*als franczoschisch*) and Italian (*wälsch*). While from the text and picture there doesn't seem to be any difference between an Italian and typical low thrust made with a "sword" grip, the French thrust to the breast is indeed completely different and not encountered in any other source from that period known to me.

[20] BF, f 22r.
[21] BF, f 26r.

3 Augsburg Dagger Teachings in Later Sources

BF on its own would be "just another fight book", but tracing the recorded techniques of the Augsburg Group in later sources reveals an interesting picture. These are:

- *Berlin Picture Book* (BP), 1510s
- *Albrecht Dürer* (AD), 1512
- *Gregor Erhart* (GE), 1533
- *Goliath* (G), 1535–40
- *Paul Hektor Mair-Dresden* (PMD), 1542
- *Anton Rast* (AR), 1553
- *Lienhart Sollinger-München* (JWM2), 1556

Let's examine these manuscripts in chronological order and compare them.

3.1 *Berlin Picture Book* (BP), 1510s

This is an extremely interesting source. It contains a large number (118) of illustrated dagger techniques without text, of which at least 20 are present in other manuscripts from Augsburg tradition. The style of depictions is reminiscent of other works made in Albrecht Dürer's workshop, though this connection has never been formally established.

Almost all illustrations present in BF, including four that are otherwise-unique to that source, are contained in the BP, and a significant number of otherwise-unique techniques from AR are also present, including the full set dealing with being grabbed by the collar (plus one additional technique unique to BP). Several unique techniques from G are missing, though. Most of the time the illustrator in BP seems to be choosing the same or similar key technique moment as in the other sources, including a very complex drawing on 41r-A which is the same as in AR on 42v (Fig. 4). Also, while striving for anatomical correctness, there seems to be some awkwardness and possibly several mistakes in depicting weapon and combatant positions in relation to one another, such as an impossibly twisted joint on folio 47r-B or a possible dagger painted in front of the forearm on 58v-B, which make precise identification of techniques difficult and conclusions uncertain.

3.2 *Albrecht Dürer* (AD), 1512

The association of AD with material presented in BF has been widely studied.[22] However, while we can without question see the adoption of wrestling, sword, and Messer techniques, it is not so with dagger. There are two pictures with techniques

[22] Dörnhöffer 1907/1909, Welle 2014, Hagedorn 2021.

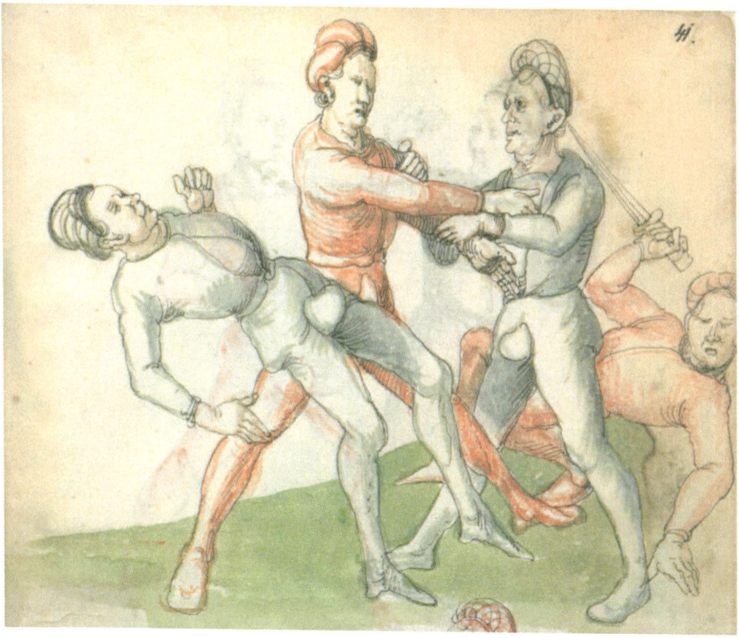

Figure 4: Technique against multiple opponents (top: BP, f 41r-A; bottom: AR, f 42v).

Figure 5: Crossing the legs (top: AD, f 66v; bottom: G, f 90r).

found in G and AR, but no real connection with BF. The crossed legs on folio 66v (number 15; Fig. 5) are similar to the depiction of this technique in G and are not typical to other Medieval sources.

Overall, as far as dagger goes, AD seems to play only marginal role in preservation and propagation of the teachings and point to inspiration from another source.

3.3 *Gregor Erhart* (GE), 1533

This treatise contains text describing seven dagger techniques attributed to Jobst von Württemberg on ff 178r–179r. They can all be shown to be textual variants of material present in BF, G, and AR, with all reference to pictures omitted (Table 1).

It's not clear if it was Jobst or Erhart himself who decided to forego pictures and remove references from the text. Erhart does include another set of dagger teachings in his treatise, which also omit picture references present in other sources, so it's possible that it was him and original Jobst teachings did follow the illustration with description pattern, but there is no way to be sure.

Apart from missing reference to pictures, the text also has no technical terms such as names for the thrusts or injections from Liechtenauer's teachings. There's little that we can deduce from the order or selection of these particular techniques. We don't know if attribution to Jobst denotes the name of the tradition originator or just one of his students. The name of the town tells us even less about these teachings or their proliferation. The only thing that we can say for certain is that neither BF nor G could have been the sole sources for these teachings.

With Bartosz Starko, I also posit the existence of an earlier manuscript (proto-Erhart) of which this one is a direct copy, based on our analysis of other material contained therein.[23] Considering that Jobst's teachings are copied verbatim in JWM2, that shares the same source, so ordering and all other changes must have been introduced in proto-Erhart manuscript.

Table 1: Plays in GE compared to BF, G, and AR.

GE	BF	G	AR
1	22r	94r	-
2	28r	96r	-
3	22v	-	36r
4	23r	92v	36v
5	-	93r	37r
6	24r	94v	38r
7	24v	95r	38v

[23] Walczak & Starko 2018.

3.4 *GOLIATH* (G), 1535–40

Dagger teachings associated with the Augsburg group are an oddity in this manuscript, which is mostly a direct copy of PD with occasional illustrations. Presented without attribution, they follow the "picture with description" format—something atypical for Goliath as well—and consist of seventeen techniques in total, of which one does not have a caption. Four of them are present in both BF and AR, four only in BF, five only in AR, and four are unique to G and don't even appear as pictures in BP.

The title page of this "book of dagger" suggests that the techniques should be done in armor (*Im pantzer*), however all illustrations show combatants without it. Daggers in G are also shorter than the ones depicted in other sources belonging to this tradition. Folio 90r shows crossing of the legs associated with the 16th century material, uncharacteristic to what we can encounter in the material from 15th century and earlier (Fig. 5). Pictures also occasionally display signs of being corrected after the initial sketch and maybe even inking, but it is only clearly visible in person, not nearly so on digital reproductions.

The text removes both names for thrusts associated with nationalities (misreading Italian as "false"), but adds another technical term—*die Waage*—present in BF (but only in the sword teachings).

3.5 *PAUL HEKTOR MAIR-DRESDEN* (PMD), 1542

Paul Hektor Mair authored three separate versions of his magnum opus in the years 1542–56, slightly changing its content with each edition, but I've decided to only include the earliest one as an example, as the changes are not relevant to the current discussion.

From historical data and his notes inside the BF, we know that Mair most likely had access to it in its current form. For his treatise, he commissioned redrawing and adapted eight out of thirteen techniques from part C and twelve out of fourteen from part A. As mentioned before in description of part C, his text has little to do with the original, as evidenced by the following example: [24]

BF – 26v	**PMD – 11r**
	Ein prust stosz mit ainem Gemecht stosz
Item sticht dir ainer oben zu dem gesicht so slach im mit deiner tencken hant aussen auf sein rechten hant so sticht er sich selb in dy hoden vnd stich inn dann mit deinem degen oben ein zu dem hals	*Item schickh dich allso Mit disem zufechten. stand mit deinem lincken fuosz vor. den daumen auf deinem knopf vnd stosz Im nach seinen prust. stost er dir also zu. so nimb Im das ab mit deiner dolchen klingen vnd stosz Im nach seinen gemechten. wirstu des gemecht stosz*

[24] All transcriptions and translations in this paper were made by the author.

als hie gemalet stet[25]

gewar. so fall Im mit deiner lincken hand nach seiner rechten wend damit seinen tolchen von dir.[26]
[... – Mair's text goes on]

I am going to forego including an exact concordance table for the sake of space and because it is irrelevant to the subject of this article. It is obvious that Mair, despite being an avid collector of fight books, only saw utilitarian value in the pictures of BF dagger and mostly ignored its actual content.

3.6 *Anton Rast* (AR), 1553

The introductory page of this manuscript claims that it was created (or commissioned) by Paul Hektor Mair based on the book made by master Anton Rast after Rast's death in 1549. The original manuscript is not known to have survived. The Augsburg Group teachings for all weapons are placed between gloss of Liechtenauer's long sword and Rast's own *Kampffechten*. They follow the typical "picture with description" fashion and are part of the organized progression from wrestling through dagger and Messer to long sword, with transitional techniques between various parts (dagger vs. Messer and Messer vs. long sword). This organization makes better sense than in other sources and shows that the Augsburg Group teachings—at least for Rast—constituted an inseparable whole. It's not clear if this was Rast's own addition to this material, something that was passed on to him, or a decision by Paul Hektor Mair.

There are fifteen dagger and two dagger vs. Messer techniques. Five techniques present only in G have almost the same order in AR, two are unique to AR and BF, four are common to all three, and four are unique to AR alone. The pictures are stylized 16th century renderings of the 15th century clothing and weapons, and in composition are quite similar to those found in the BF.

The text is straightforward, there are no special technical terms in the dagger section.

If we are to believe the words of Paul Hektor Mair, Anton Rast was 70 years old when he died in 1549. This would make it possible for him to receive his first fighting instructions at the end of the 15th century, perhaps even to be a direct student of the originator of the Augsburg tradition, which could explain the

[25] "Next: when one stabs you towards the face, then hit his right hand with your left hand from the outside, so that he stabs himself in the manhood. Then stab him with your dagger from above towards his neck, as it is painted above."

[26] "A breast strike with a strike towards manhood.
"Next, prepare yourself for combat like so: stand with your left foot forward and thrust him in his breast. If he thrusts you as well, then grab that [thrust] with the blade of your dagger and thrust him in his manhood. If you want to defend yourself from this strike, then drop your left hand on his right one and there twist his dagger away from yourself..."

extreme similarity of his alleged book to the material present in BF and the stylization attempted by Mair.

It's worth noting that none of Rast's teachings in AR made it into the final edition of Paul Hektor Mair's magnum opus in any way. It looks like when Mair decided upon its contents in early 1540s he did not have access to this manuscript, and he chose not to incorporate them in his later editions even after he had such an opportunity.

3.7 *Lienhart Sollinger-München* (JWM2), 1556

On ff 66r–69r, this treatise contains teachings ascribed to Jobst von Württemberg which are almost an exact copy of those found in GE. However, as we have suggested in another paper, they are most likely both independent copies of an earlier manuscript.[27] The text doesn't contain any meaningful changes and there are no images.

4 Overall picture

Let's examine the full concordance table which takes into account the most important manuscripts: BF, G, and AR (which form the "core" of the teachings), as well as GE and BP. Because the teachings are scattered, incomplete, and their order is mixed up, no source immediately jumps out as the candidate for the base. I decided to assign this role to AR, mostly because it looks to be the best organized from them all, but the same could be done for G or BF without any noticeable difference.

In Table 2, I marked techniques unique to one of the three core sources in *italic* on grey background, while those also absent in BP in ***bold italics***. Additional techniques present only in BP are omitted for the sake of space. It's possible that despite my best efforts to find equivalents in BP, I missed some of them.

The overall picture that emerges from this concordance is the clear lack of base. None of the manuscripts seem to be obviously incomplete or have reordered quires, there are areas of continuity, and the gaps between some techniques would often fall between recto and verso of folios. At first glance it would support the thesis of dagger material of that period being mostly unsystematized and opportunistic and semi-random selection of material by all authors.

It might also suggest that if there was a common source, it could have been just a bunch of single-sided unbound folia, perhaps a training/memory aid, which could be freely reordered or maybe even updated with additional techniques without leaving any trace of such changes. While I know of no such artifact surviving, and both single-side painting and separated pages would be an oddity then, Scheller in his seminal book on artistic transmission points out several instances of

[27] Walczak & Starko 2018.

BF	G	AR	GE	BP
—	91r	33r	—	41v-AB
—	91v	33v	—	41r-B
—	90r	34r	—	55r-B
—	90v	34v	—	44v-B (?)
—	—	*35r*	—	64v-B
28v	—	35v	—	48r-B
22v	—	36r	3	64r-A
23r	92v	36v	4	G: 55r or 61v-B; others: 38r-A, 39v-B or 51r-B–52v-A
—	93r	37r	5	50r-B
23v	93v	37v	-	55v-B
24r	94v	38r	6	37r-A
24v	95r	38v	7	37v-A
—	—	*39r*	—	41r-A
—	—	***39v***	—	—
—	—	***40r***	—	—
22r	94r	—	1	37r-B
25r	—	—	—	43r-A
25v	89v	—	—	64v-A
26r	—	—	—	47r-B
26v	—	—	—	38r-B
27r	96v	—	—	—
27v	—	—	—	39r-A
28r	96r	—	2	40r-B
—	***92r***	—	—	—
—	***95v***	—	—	—
—	***97r***	—	—	—
—	***97v***	—	—	—

Table 2: Dagger techniques of the Augsburg group.

separated bifolia being handed to another artist.[28] The proliferation of paper in the 15th century Germany and its civic usage for everyday activities also speak to the possibility for such an oddity to exist.

The alternative hypothesis would be a conscious reordering of techniques by each author, while preserving the key features and sometimes even the style of illustrations. It's not implausible either, considering the disjointed nature of dagger teachings overall, but sounds less probable when more sources are considered. Contrary

[28] Scheller & Hoyle 1995: 20.

to this, sword and Messer sections from the Augsburg Group follow a consistent order in all manuscripts wherever they are copied.

This contrast on its own raises a lot of questions to which we don't really have an answer. Has there ever been a core that was transmitted? When were the techniques only present in AR and G, such as dealing with being grabbed by the collar, added? How did these techniques make their way into BP with the collar group including an additional technique? How do we explain occasional large consistency of weapon, keyframe selection, perspective, and depiction between BF, AR, and BP? Why did the author of G choose to only include the dagger section in his treatise?

5 Comparing techniques

In the hope of having some of those and other questions answered, let's make some more detailed comparison of three techniques present in all five major manuscripts.

BF 24v:

> *Item sticht dir ainer oben ein zu dem gesicht so vass deinen degen in paid hent als vor und vach den stich darauf und lauf vast hinder sich zuruck mit im und dauch ymen [ihm] vber seinen kopf an dem hals als hie gemalt stet so zeuchstu in auf dy erd*
>
> Next, when one stabs you from above towards the face, then grasp your dagger in both hands as before, catch this stab with it, run surely backwards with him, and plunge over his head on his neck as it is painted here, thus you will pull him on the ground.

G 95r:

> *Aber ein Stuckh.*
>
> *Sticht dir einer, oben starckh ein nachdem an= gesicht so winnt vff mit gewapenter hannt alß fur vnnd gee vast Inndeß, Vnnd ruckh im den tegen, hintersich vff, Vber sein heubt vnnd zuckh in starckh fursich, Gegn der erden, Vts.*[29]

[29] "Vts." is an obvious reference to the picture, though WELLE decodes it as *ut supra*, "as above" (WELLE 2017: 80).

> Another piece.
>
> When one stabs you from above with strength towards the face, then turn upwards with armored hand as before, go surely at the same time, jerk away his dagger backwards over his hood, and pull him strongly towards you, to the ground. [As painted]

AR 38v:

> 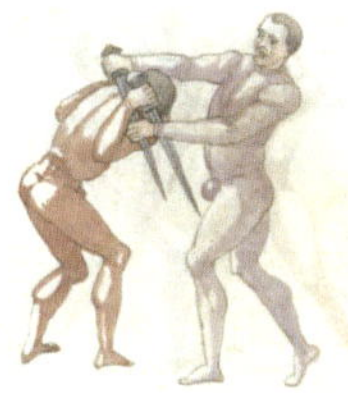
>
> *Item Sticht dir ainer nach dem halß. So var aůff mit ge= wappneter hand mit dem Degen. Vnnd vach den Stich daraůff. Vnnd drůckh Im seinen Degen vber das haůpt. vnnd Zeůch in nider zů Erden. Als da gemalet steet.*
>
> Next, when one stabs you at your neck, then go up with armored hand with the dagger, catch the stab with it, push him his dagger over his head and pull him downwards to the ground, as painted.

GE 179r and BP 37v-A:

> 
>
> *Item sticht dir ainer oben nach dem gesicht so wind auͣf als vor vnder sein degen vnd laůf vast in in vnd taůch in den degen vber sein haůbt an den hals vnd zeůch in zů der erden*
>
> Next, when one stabs at you from above towards the face, then turn upwards as before below his dagger, run surely into him, plunge him the dagger over his head on the neck, and pull him to the ground.

All sources describe the same technique, but using different phrases and sometimes even introducing concepts from other traditions or parts dealing with different weapons (e.g. *Indes* in G). Even the reference to the picture drawn is placed differently inside the text. However, there is no denying that this is the same move, which is also evidenced by virtually the same picture in BF, BP and AR. Only G shows the opposite leg forward, which we will get back to later.

This technique is not a common one and could be considered a signature of the Augsburg Group. To have it preserved in so many sources seemingly independently of one another suggests some form of communication and memorization that did not rely on a written word. It's possible that in this case the BF picture was used as a base for BP and AR or that they all were based on the same source which has not survived. Goliath, on the other hand, does not seem to share this lineage.

Considering a significant time difference between dating of the first and the last of them, some sort of intermediary must have existed for the technique to be transmitted that was able to preserve its key movements.

Let's examine another technique which contains an interesting divergence, though most of the text probably stems from the same original one. I have marked in bold the most meaningful difference between the texts.

BF 24r:

> 
>
> *Item sticht dir ainer oben ein zu dem gesicht so vass deinen degen in paid hent und vach den stich dar auf und* ***wint im den knopf*** *uber seinen rechten armen als hie gemalt stet so slechstu im seinen degen in das angesicht*
>
> Next, when one stabs you from above towards the face, then grasp your dagger in both hands, catch the stab with it, and turn the pommel over his right arm as it is painted, thus you will strike his dagger into his face.

G 94v:

> 
>
> *Aber Ein Stuckh:*
>
> *Ob dir einer gar gern nachdeinem ansicht, so fasß deinen Degen zu Gewappenter hannt, Vnd gee hoch auf, Vnnder seinen stich, Vnnd* ***wint im mit deinem knopff, oder spitz*** *vber seinen arm, Als do gemalt stet, so schlechstu im seinen degen innß angesicht, Vnnd nimbst im den Degen*
>
> Yet another piece:
>
> If one eagerly endangers your face, then grasp your dagger in the armored hand, go high up under his stab and turn your pommel or point over his arm, as it is painted, thus you will strike his dagger in his face and take his dagger away.

AR 38r:

> 
>
> *Item Sticht dir ainer oben nach dem halß. So vaß deinenn Degen zů gewappneter handt. vnnd* ***wind Im deinen Spitz.*** *vber seinen arm. So prichstů Im den Degen aůß. als da gemalet steet.*
>
> Next, when one stabs at you from above at the neck, then grasp your dagger in the armored hand and turn your point over his arm. Thus you will force the dagger [from his hand] as it is painted.

GE 178v–179r and BP 37r:

> *Item sticht dir ainer oben nach dem gesicht so vas dein degen in paid hend vnd vach im den stich aůf vnd taůch in wol zů rück vnd* ***wind im mit deinem spitʒ*** *vber sein arm° so brichstů im den degen aůs duͣ magst in auͣch* ***mit deinem knopf vber winden*** *vnd auͣf brechen*
>
> Next, when one stabs you from above towards the face, then grasp your dagger in both hands, catch his stab above, dip him surely backwards, and turn your point over his arm, thus you will force the dagger [from his hand]. You may also perform the turning above with the pommel and disarm him.

Had we not had G, we could have separated these as two different techniques which use the same defense against the same attack and possibly have two different outcomes (disarm or thrusting opponent's dagger into his face). But G merges them into a single description, even though its illustration appears to show the variant where it's the point that gets twisted over the opponent's arm. BP actually seems to illustrate both variations on folio 37r, repeating BF's possible mistake on the "point" variant.[30] GE adds the pommel twist as an afterthought, while BF records only the pommel twist version and AR uses just the point.

We can also see another injection from a different tradition: this time it's Liechtenauer's *Kampffechten*, where the term *gewappneter hant* is used in reference to a sword being held with the left hand in the middle of the blade to give it better thrusting capability against armor. Both AR and G make use of this term, while BF and GE don't.

The picture in G is unique again, both in its style and execution, while pictures in other sources seem pretty much consistent. Analysis of descriptions also shows that the core of this technique remains the same, regardless of the variation in wording and syntax. As far as chronology is concerned, it's impossible for us to establish whether the "point" variant was known at the time when BF was created or if it was invented later during experimentation with the original "pommel" variant.

Let's continue now with a potentially similar divergence that could have happened through a possible misreading or misunderstanding. I marked the text where a possible confusion could have happened in bold.

[30] The defender's dagger pommel should be behind the opponent's arm, not in front or above it, as it can be seen in BP. BF has this pommel awkwardly painted, as if someone added it after the fact and it had originally been (correctly) hidden behind the forearm.

BF 23r

> *Item sticht dir ainer zu dem gesicht so* ***nym den degen auf deinen rechten armen und vach den stich dar auf*** *und wint im mit deinen spicz uber seinen armen und zeuch an dich als hie gemalt stet so nymstu im den degen oder du prichst im den armen ab*
>
> Next, when one stabs you towards the face, then grasp your dagger on your right arm, catch his stab with it, turn your point over his arm and pull towards yourself as it is painted here, thus you will take his dagger away or break his arm.

G 92v

> *Item sticht dir einer oben nach dem angesicht nach mit gefaussten degen, so* ***winnt mit deinem degen ebichen vff, Vmbwint sein degen,*** *vnd vmbwinnt im deinem spitz, yber sein rechtn arm, Vnnd reiß an dich, vts, So nimbst im den Degen aus der hant.*
>
> Next, when one stabs you from above towards the face with the dagger in his fist, then turn your dagger inverted upwards, turn his dagger around, turn your point around over his right arm, and jerk towards yourself [as painted], thus you will take his dagger from his hand.

AR 36v

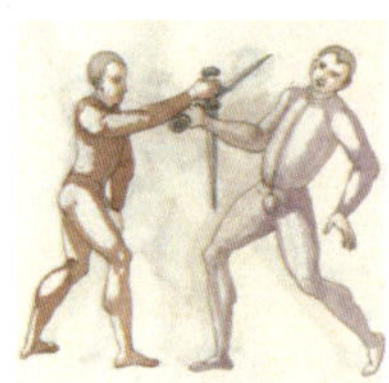

> *Item Sticht dir ainer zů, nach dem halß. So* ***wind Im vff mit deinem abichen Degen. vnden an seinen Degen,*** *vnd vall Im mit dem Spitz v̈ber seinen arm. So reibstů Im den tegen aůs der Hanndt. als da gemalet steet.*
>
> Next, when one stabs you at your neck, then turn your inverted dagger upwards, below his dagger, and drop your point over his arm. Thus you will rip his dagger out of his hand, as it is painted.

GE 178v

> *Item sticht dir ainen oben nach dem gesicht so* ***wind aůf mit deinem degen vnden an sein degen*** *vnd taůch in vast zů růck vnd wind im mit deinem spitz vber sein arm° so brichstů im den degen aůf*
>
> Next, when one stabs at you from above towards the face, then turn your dagger upwards under his dagger, dip him surely backwards, and turn your point over his arm, thus you force his dagger [from his hand].

Let's start with the picture this time. It's no surprise that BF and AR are almost the same. G again seems to show a completely different, maybe even unrelated technique. As far as BP goes, it's extremely difficult to find illustrations that we can say for certain references this technique. G's variant possibly can be found in the sequence on folia 55r or 61v-B. The other variant might be illustrated on folio 38r-A (with the defender's dagger mistakenly drawn in front of the opponent's), 39v-B (the final moment of the technique) or in the sequence 51r-B–52v-A (view from the opposite side with the dagger point not yet over the opponent's arm).

If we just examine the text from BF without referencing the picture, especially without understanding what the "grasp your dagger on your right arm" phrase exactly means,[31] it's possible to arrive at the interpretation present in G—that you defend with the dagger point already above the opponent's arm and then attempt to twist it to the right side and hope for a disarm or possibly an elbow/arm break. However, a reverse engineering of the image in BF and AR, as well as attentive reading of the description, reveals that the dagger blade should lie on the forearm instead of being held at roughly 90 degrees, and that the disarm is actually more complex, much more subtle, and turns out to be yet another signature technique of the Augsburg Group.

In this light, even re-reading the G text with more attention shows that the picture is not a valid representation of the description. If we also take into account numerous corrections of the G pictures (unfortunately only obvious when looking at a physical copy of the manuscript), we can hypothesise that the artist could have been working without the author's supervision, didn't have exposure to the teachings and lacked author's feedback, or that there was no will to correct some of the inconsistencies and mistakes early. On top of it all, considering how big the error in this technique is, maybe correcting this particular picture would have required redoing the whole quire and so it was left as it is.

6 Discussion

6.1 Oral literature hypothesis

From the above examples it should be clear that while the text in each manuscript does describe the same techniques and contains several common phrases, such as *item sticht dir ainer*,[32] the exact descriptions are different enough that it's impossible to trace it using traditional textological analysis. It's no wonder that WELLE was unable to find the relationship between them and did not pursue the subject further.

However, this situation is eerily reminiscent of what REICHL writes about oral medieval literature.[33] The textual variations confuse experienced editors and

[31] *nym den degen auf deinen rechten armen* (BF, f 25v).

[32] "Likewise, if someone stabs you..."

[33] REICHL 2016: 22.

prevent establishing the stemma. These variations could easily have resulted from committing text to memory, while its independent recall and writing down occurred at a later date by people who were familiar with the original meaning but used their own words to explain it.

Considering, that the accompanying pictures are extremely similar in BF and AR (G is slightly different in this case) I posit here another hypothesis, that this is an example of such an oral tradition and it's being supported not by any text, but solely by images and bodily knowledge of students, similarly to what we can see in part C of the BF. Analysis of oral literature has shown that it was very stable in terms of preserving core message[34] and that writing does not need to be the only memory prop. The role of images for both memorization and recollection has also been described in more detail by WENZEL in his work on memory devices[35] and raised briefly by MÜLLER.[36]

I can add to this my personal bodily experience obtained while attempting to reconstruct these and similar techniques. Identification and recall of a previously learnt technique from a single depiction of its key or signature moment is quite effortless, even after significant time has passed from the original practice. Of course, it is only possible because I had been able to deconstruct and translate the original text into the working movement beforehand and learnt it through repeated practice, which caused my body to internalize the movement patterns. Regardless of the validity of my actual execution of the recorded technique, this involuntary embodied knowledge experiment resulted in a possible confirmation of the identification and recall hypothesis.

It seems also plausible that when G manuscript was being created, its author committed his own recollection of the teachings to paper, which was then interpreted again by the artists and models and resulted in several mistakes that either were missed or deemed not worthy of correction afterwards. This would then make it an interesting case of how replicating written transmission is difficult even when the skill is still alive and relatively common.

6.2 "STEMMA" ANALYSIS

I'm going to join the choir of other scholars who say that it's extremely difficult to speculate about the exact relationship of all five base sources, especially when dagger is considered. I will go even further and claim that—given all what we mentioned before—it's not implausible for the dagger teachings to have originated from a different source than the sword or Messer and been added to the corpus later.

Based on available evidence, there are several hypotheses that we can make.

There seems to have existed a common pool of dagger techniques from which all authors picked their pieces. Original teachings could have been a collection of loose

---

[34] Ibid: 27.

[35] WENZEL 1995.

[36] MÜLLER 2016: 308.

Figure 9: Three dagger plays (AD, f 67r).

single-sided pages with illustrations without any text. These could have been freely reordered when needed and new techniques could have been inserted anywhere at any point in time as long as such collection existed. This would explain weird discrepancies in ordering of the pieces in various manuscripts and variations in descriptions. The authors of BF, G, the original AR, and possibly the original Jobst seem to have had access to this source.

AR and BF seem to be related most closely, both when it comes to text and pictures, despite the breadth of time that divides them and uncertainty about Paul Hektor Mair's interference. However, the pictures in AR are clearly stylized for the 15th century, which is very atypical for Paul Hektor Mair and that seems to suggest his attempt to faithfully replicate the original. It's something that he had done with some other purely textual sources as well. AR does seem to incorporate a number of techniques which form their own group and were not part of BF, but this could just be the original collection being altered sometime after BF was created.

The author of G seems to have had access to the same collection that AR had and possibly added a few more techniques of their own, since they are not present in any other source from this tradition. In other parts of G, there are similar additions that go beyond just copying material from other sources, so this would fit the pattern.

Jobst's teachings, as recorded in GE, seem to be mostly related to BF and AR. The order and wording make a strong argument for the existence of the original base collection.

BP incorporates only the pictures of techniques from the already-altered collection upon which AR and G were based. It isn't a very faithful copy, contains several variations and re-renderings, and also mixes Augsburg Group dagger teachings with other techniques, including ones possibly coming from the earliest known fight book of Hans Talhoffer (HTG). It's extremely doubtful that it could have been an inspiration to G or any other manuscript in question, though it's definitely possible that it contains techniques present in the original collection and absent in all other extant sources.

Mair and Dürer are peripheral at best. Mair merely takes BF images and disregards their descriptions altogether. And when it comes to dagger, AD seems to have little or no connection to the original material (Fig. 9), as opposed to other teachings associated with the Augsburg group.

Illustrating these relationships would post a significant challenge and probably bring more confusion, therefore I decided not to attempt this task.

7 Conclusions

At first glance, the dagger teachings in the BF seem to be typical for their period and region: an unsystematized, opportunistic gathering of recipes on how to deal with an opponent wielding such a weapon. However, deeper analysis and comparison to other sources belonging to this lineage reveals a case of transmission that is not purely textual, but most likely an example of an oral literature supported by pictures and bodily knowledge that was later independently put into writing by at least four different individuals.

The resulting divergence in textual descriptions, while being a bane of paleographers, is extremely useful for reconstruction. Each author describes movements in a slightly different manner, stressing different things. While this might lead to misunderstandings, such as the one in the mentioned disarm in G, it's also possible that additional details help us to better home in on the expected execution and to extract the core of the technique with more certainty, than is the case with a single text repeated many times.

Other artifacts, such as partial reordering of techniques and insertion of new groups in later sources, suggest that the original source might also not be a bound manuscript, as is the case with all other fight books that we are familiar with, but rather a collection of loose pages. This hypothesis requires more research and might turn out incorrect, but for now it seems to be the simplest explanation of the structure of surviving manuscripts.

Hye merck ain gutz stuck kumt dir ainer zu von
oben nider so mach den slag auf den mayst als war
und gee mit dem lincken fuss in den und stoss im
mit deiner lincken hant sein rechte ander zu
der erd und kum im uber den kopff als hie gestet

MESSERFECHTEN FROM THE AUGSBURG TRADITION

Lucien Lee
Bastion HEMA

There are many manuscripts detailing the how-to of sword combat on horseback and on foot, both with and without armour. In modern studies, while the art of fencing with the two-handed sword attracts the largest following, it should not be considered the sole art of combat. Alongside the texts on the two-handed sword, there also exist similar treatises on single-edged and single-handed weapons that were popular in the Holy Roman Empire during the late-Middle Ages. These weapons are known as *Messer* (or "knives").

Johannes Lecküchner wrote the largest of these Messer treatises, found in *Heidelberg* (LH, ca. 1478) and *Munich* (LM, 1482). Furthermore, other manuscripts also reproduced and included Lecküchner's material to varying degrees. Examples include *Hans von Speyer* (HS, 1491) and *Gregor Erhart* (GE, 1533). Lecküchner's works have many ideas, teaching structures, and vocabulary in common with the writings of Johannes Liechtenauer. However, the similarities do not mean that they share the same origin. Rather, these were two separate but similar works that people might collect or study. Other masters have written treatises comparing and combining elements from both works: for example, a Master Andreas compared terminology for *Messerfechten* and *Blossfechten* in 1491, and Peter Falkner, Captain of the *Marxbrüder* fencing guild, wrote a treatise in ca. 1495 (PF) that clearly took inspiration from both Liechtenauer and Lecküchner. It is evident that Lecküchner and his works were well-regarded and were copied several times over the following century.

Lecküchner's work, however, was not the only treatise written for Messer. There also exists another body of work written within 80 years of Lecküchner's own work. These treatises do not use the same structure or vocabulary as Lecküchner, and they also feature a smaller list of techniques, elaborated upon for a variety of situations.

Today, this group of manuscripts is often called the 'Nuremburg Tradition'. It is a bit of a misnomer, as the selected manuscripts seem to originate from the Swabian or Augsburg region rather than Nuremburg. While Lecküchner himself was a resident of Nuremburg, he did not seem to be largely influenced by, nor had distinct influence on, this separate tradition. In contrast, Hans Talhoffer seemed to show a much stronger connection with an Augsburg tradition in regard to his own Messer teachings.

The purpose of the paper is to build a case for the Augsburg tradition of Messer fencing. While it is by no means unique in its techniques, the particular grouping of techniques in the Augsburg tradition is special from a martial standpoint. It allows a good base for Messer combat without needing to overload with too much information, especially for a beginner. This paper will examine the extent of the martial connection between the Augsburg tradition and other masters, and propose that studying the Augsburg tradition makes for a great steppingstone to examining Leckűchner's treatise. Additionally, the analysis will provide a basic framework to understand Messer fighting outside of Leckűchner's methodology. It will allow the practitioner to tackle Messer in Leckűchner's work in contrast with the Augsburg tradition.

1 Texts of the Augsburg Tradition

The selected treatises for the Augsburg tradition are largely based on their similarities of techniques within a time period of about 80 years.[1] This means that despite spanning across a substantial amount of time, the techniques were described in similar form, from author to author. It is possible that each author might have copied from each preceding manuscript and then added their own spin on the subject, but it seems unlikely that wholesale copying was done. The only large lifting appears to have been when Anton Rast took the Messer section of *Bauman's Fight Book*, though he also modified the text to emphasize the finer details of the techniques.

It is more likely that the manuscripts examined in this paper indicate a shared pool of knowledge that was used and taught among a group of people through a period of time. Each author wrote their treatise to reflect what they considered important, and excluded different aspects.

The texts that have been examined are as follows:

Main Augsburg Tradition texts:

1) *Bauman's Fight Book* (BF), some parts likely from ca. 1470s (compiled in 1556), Southern Germany.
2) *Glasgow Fight Book* (GF), likely from ca. 1508 Holy Roman Empire.
3) *Anton Rast* (AR), likely from ca. 1553 Augsburg.
4) Jobst von Württemberg in *Jörg Wilhalm-München #2* (JWM2), likely from ca. 1556 Augsburg.

Secondary texts:

1) *Andre Paurenfeyndt* (AP), likely from ca. 1516 Vienna.
2) *Paul Kal-München* (PKM), likely from ca. 1470 Bavaria-Landshut.
3) *Hans Talhoffer-München* (HTM), likely from ca. 1467 Württemberg.
4) *Peter Falkner* (PF), likely from ca. 1495 Augsburg.
5) *Hans Leckűchner-München* (LM), likely from ca. 1482 Nuremburg.

In examining the above manuscripts, this paper has chosen four main sources (BF, GF, AR, JWM2) to represent the Augsburg tradition. These texts have shown strong similitude and hence are grouped together. They are selected for their

[1] From the 1470s to the 1550s.

emphasis of certain technique sets, and for their lack of Leckùchner- and Liechtenauer-specific vocabulary when describing Messer techniques. The selected texts had very limited to zero use of vocabulary such as *Zornhaw*, *Winker*, or *Bastei* to describe movements and positioning. They tend to describe the movements directly and simply, i.e., "place your feet here and put the Messer there". Where terminology is used, it tends not to be the same one used in the Leckùchner and Liechtenauer traditions.[2] In regard to the secondary texts, they are treatises that will be used to observe certain peculiarities, comparisons, or similarities, and will be looked at during the analysis and comparison section.

2 TECHNIQUES OF THE AUGSBURG TRADITION

The following chapter will examine the common thread of the Augsburg tradition which will be introduced as 'technique sets'. This paper will not be discussing any particular passages, per se, but rather the aggregated understanding of the technique. The aim is to examine the technique grouping holistically, so that one can understand what the limitations and uses of a particular set of movements are.

Each technique set will be examined for its specifics in order to understand how widespread they are within the Augsburg tradition. The utility of the techniques will also be studied from a martial viewpoint, to understand if the particular persistent grouping of these technique sets is purposeful.

What exactly is a 'technique set'? For the purpose of this paper, it is a collection of variations of techniques that all react to various stimuli with responses that use similar (if not identical) positions and/or movements. These responses might be implemented with slightly different details depending on the precise situation, but the core idea of the response will remain consistent throughout each example of the technique. The technique sets are not limited to any single description of a technique but encompass groupings of passages that all have a commonality in their movements.

This paper identifies four main technique sets and two secondary sets from the Augsburg tradition texts. Most of the techniques fall within these four main sets, with only ten out of 63 paragraphs falling into the secondary sets. The ten passages which fall within these secondary technique sets are separated into *langen Versetzen* and others. They will not be the subject of this paper, as nine of these passages are found only in JWM2 and one in AR. Given that examples of these ten passages are only in the manuscript that they are found in, it would be hard pressed to say that they have a crucial part in the Augsburg tradition. Thus, this paper will focus on the four main sets.

[2] There are exceptions to this. They might describe positions which is common to both traditions, such as *vom Tag*. However, these are far and few in between.

Figure 1: The direct displacement according to Hans Talhoffer (HTM, f 113v).

The main four sets are:

- *Schlechten Versatzung*, or "direct displacement"
- *Verkerter Hant*, or "turned hand"
- *Gewappnet Hant*, or "armed hand"
- *Mitelhaw*, or "middle hew"

The following section will be detailing out the technique sets: they are listed with their name and translation, with alternative translations offered in the footnotes. The description of each technique set will cover details such as how they are to begin, what they would be used to defend against, and the Messer and foot movements. This is to provide a clear idea of what the dominant theme of each one is.

2.1 Technique Set 1: *Schlechten Versatzung*, or "Direct Displacement"[3]

The *schlechten Versatzung* is described as beginning with the point down or to the side, with the edge and hilt above, and taking the blow onto the edge of the Messer

[3] Other translations include "direct displacement" and "simple parry".

(Fig. 1). A variation of this technique can be seen as the *zoll Versatzung*,[4] where one hangs the blade (with the point down) to ward off a thrust. Both movements are the same in the way they move to oppose the opponent's attack.

The starting position is usually described with the Messer at the hip. The left foot is placed forward, and the Messer is above the right leg by the side of the hip. The Messer edge is facing down, and the point is facing the opponent.

The movement is described as a winding action from beneath, likely turning the Messer from an edge-down to an edge-up position. As this happens, the point starts to point downwards, and the Messer is then moved towards its ending position.

The parrying position is where the Messer moves to meet the opponents blade. The Messer extends out towards the incoming attack and meets it with the edge. This attack can be a thrust or hew from above, or a thrust from the front.

The ending position is where the Messer successfully displaces the opponent's blade, either by directly blocking a downward hew, or by shifting aside a thrust. The opponent can either strike and not bind, or bind upon the defender's Messer upon striking. Either situation is accounted for in the passages.

2.2 TECHNIQUE SET 2: *VERKERTER HANDT*, OR "TURNED HAND"[5]

The *verkerter Handt* is described as displacing a blow by turning the Messer to intercept the incoming blow with the *Nagel* ("nail") or on the flat. The hand and wrist are turned, with the knuckles facing the side of the incoming blow (Fig. 2).

The starting position of this technique is usually described with the Messer next to the knee, pointed down to the ground, in *weschel* ("change"). At times, it is also described as "lying by the leg", with both right- and left-foot-forward passages written as well.

The movement is described as shooting forward and upwards with the point, or to drive with the point, turning the hand as the Messer is brought up.

The parrying position is where the Messer meets the opponent's blade with the flat or the *Nagel*. The forearm and bicep are tensed and shortened with a bend in the elbow due to the rotation of the wrist and hand. The defender should catch the opponent's hew with the arm as outstretched as is comfortably possible.

The ending position is where the Messer successfully displaces the opponent's blade, either by directly blocking with the flat and/or the blade sliding on the *Nagel*. The opponent's blade is likely stopped on the *Nagel*.

If a *Nagel* is not present on the Messer, as in the case of an arming sword, the technique can be modified to catch the opponent's blade with the back edge of the

---

[4] *Zoll* here is actually a medieval unit of measurement, akin to an inch. Other meaning for *zoll* include "customs" or "dues", "cylindrical piece, log". I am unable to think of a suitable translation, with the closest being "length displacement". "Inch displacement" doesn't reflect the technique well. Thus, I feel best to leave it as *Zoll*.

[5] Other translations include "'inverted hand" and "reversed hand".

Figure 2: The turned hand according to Paul Kal (PKM, f 72v).

Messer, and the opponent's blade should slide down towards the crossguard. A thumb grip on the sword would facilitate presenting the back edge for displacing.

2.3 Technique Set 3: *Gewappnet Hant*, or "Armed Hand"[6]

The *gewappnet Hant* is a technique that displaces a blow by holding the Messer with both hands. The left hand is described as grabbing the tip or the front of the Messer. One catches the blow with the "armed hand" directly on the Messer, and opposes the opponent's strike. This technique is described as being used against those who commit to hitting hard,[7] or who use larger and heavier weapons such as spears or swords.

[6] Another translation is "armored hand".

[7] Perhaps in the sense of *Buffel* strike, or with two hands on a Messer or sword.

The starting position of the "armed hand" is usually with the Messer in a low position. This can be done with the Messer by the side of the hips pointing at the opponent or in *wechsel.*

The movement is described as grabbing the Messer with the left hand, with the left palm facing the same direction as the right palm. The stepping should close the distance quickly, and there should be a sense of urgency in the movement. This is a fairly important point, as parrying too far away from the opponent with the "armed hand" would leave one at a distance disadvantage, especially if the opponent has a longer weapon. Take as many steps as are reasonably required to ensure that you have closed with the opponent. The follow up is typically executed with a winding to the point and shortening with the Messer; it is best to be closer to ensure that one is able to strike the opponent.

The parrying position is with the Messer held well above the head, lest a heavy blow collapses the arms. The elbows are almost fully extended, but retain a slight bent to ensure that one can absorb the shock, and to keep rotational ability of the

Figure 3: The armed hand according to BF (f 32r).

Messer. The opponent's blow is parried directly, towards the middle of the Messer. Take care not to parry the blow with either hand that is on the Messer.

The ending position is where one has closed with the opponent and has parried the heavy blow. With that, one should be in good position to wind the point to the opponent's face, and consequently push the opponent's weapon away. It is also typical to proceed with a *ringen* ("wrestling") technique (Fig. 3). If one has closed in quickly, one might also be able to catch the opponents blow before it accelerates to full power. This helps to ensure that one will not take the full brunt of the blow.

2.4 Technique Set 4: *Mitelhaw*, or "Middle Hew"[8]

The *Mitelhaw* is a technique that is named like a hew, but it seems to indicate a position where one's Messer is held on the left by the hips with the point facing the back. The movement of the hew is described as moving from the left to the right side. The hew often travels either in an arc-like motion[9] or horizontally across. This technique often instructs a hew to the opponent's arms or hands with the initial strike, as much as it describes a displacement.

The starting position is referred to as taking the "middle hew" with the right foot forward, though the left is also mentioned. The hand position can also be inferred from the phrases "from the scabbard" and "on the left". The Messer takes position with the hands by the hips on the left and under the arm (Fig. 4).[10]

As the strike descends, the movement is described as coming from above and over onto the opponent's arms or hands. In order to perform the "middle hew", one can consider the Messer hand ascending towards the shoulder before proceeding out towards the opponent. The key point is that the Messer begins at the hips but turns into an *Oberhaw* targeting the upper limbs and hands. Where thrusts are concerned, the defense from the "middle hew" position is to strike more horizontally, likely intercepting the attack at a cross. To be entirely horizontal with the incoming thrust is dangerous, as the thrust would not be displaced.

An important part of this strike is the movement of the feet. The footwork with the "middle hew" is said to be a "spring", or a "good step", as well as "out of the way". If one is not displacing the incoming attack directly, it would be prudent to be away from the path of the opponent's strike. A "good step" out of the way, typically with the left leg, is instructed along with a hew onto the hands.

The ending position is where the Messer has successfully struck the opponent's hand, thereby disarming him. Other scenarios include striking the opponent down after displacing the thrust, or applying an appropriate *ringen* technique.

[8] Other translations include "middle cut" and "middle strike".

[9] From hip to shoulder and down onto the target.

[10] This is also similar to the first ward in the *Walpurgis Fight Book* (LF).

Figure 4: Drawing the Messer according to Peter Falkner (PF, f 40r).

3 Analysis

3.1 Distribution of Passages

When scrutinizing the Augsburg tradition Messer text, one may find that most of the passages fall into above mentioned technique sets. While there are some paragraphs that do not fit, they are but a small fraction of the techniques examined.

The technique sets:

- *Schlechten Versatzung*, or "direct displacement" (all manuscripts; 16 paragraphs total)
- *Verkerter Hant*, or "turned hand" (all manuscripts; 16 paragraphs total)
- *Gewappnet Hant*, or "armed hand" (all manuscripts; 10 paragraphs total)
- *Mitelhaw*, or "middle hew" (2 manuscripts; 10 paragraphs total)

Of the main texts, JWM2 accounts for 39 out of 63 paragraphs of the techniques, approximately 62% of the material. This does not discount the validity of the other main texts, but rather only highlights it as the most comprehensive. This allows one to understand the different variations of how to use the fundamental movements in a set. JWM2 is certainly the widest in breadth, and whose material appears in all the sets.

The next most extensive manuscript is GF, which also contains paragraphs pertaining to all four main sets. There are 11 paragraphs evenly distributed through these sets.

BF has eight paragraphs that are evenly spread out across the "direct displacement", "turned hand", and "armed hand" technique sets.

Lastly, AR has six paragraphs across the main sets of "direct displacement", "turned hand", "armed hand", and in the secondary set of "others". It should be noted that AR's material, where it coincides, is extremely similar to BF.[11]

The distribution indicates that there is a strong emphasis on the "direct displacement", "turned hand", and "armed hand". While the "middle hew" is not as strongly represented within the main text examined, it comes up quite strongly in supporting texts like those by Talhoffer. Specific examples can be found in the footnotes when comparing the connection to different masters.[12]

3.2 Breakdown of Techniques

To help develop an understanding of the intention and context behind the Augsburg manuscripts, it is important to examine common threads between the Augsburg manuscripts and their passages. We will look at a breakdown of the technique sets into some common movements or actions for comparison. This allows one to scrutinize if the collection of technique is more than a gathering of knowledge or if there is an attempt to build a structure. Other key points such as the idea of "dire position", and the pattern of the passages, will also be explored so as to how it can illuminate the intended audience of the manuscript.

To start, we will break down the technique sets into several categories:

1) Starting placement of the Messer
2) Attacks that the technique set defends against
3) The footwork movement during the defence
4) The movement of the Messer during the defense
5) The *ringen* follow-up when in *ringen* range

This will allow one to understand the perceived required level of skill of the practitioner. In general, one would find that there is a conciseness about what the

---

[11] As mentioned, it is likely that Rast's material could have been based directly on that found in BF.

[12] Note that the author's own translations of the main Augsburg tradition texts are available at the end of this paper.

practitioner needs to know in order to perform the techniques. It would seem that the practitioner only needs to know a minimal amount of fundamental movements in order to perform the techniques. At the same time, each of these technique sets covers different aspects of combat, in terms of range, potential unequal situations, and common attacks from opponents.

Starting placement of the Messer

Schlechten Versatzung:	Hip, onto leg, by foot over leg, *wechsel.*
Verkerter Hant:	*Wechsel,* by foot over leg, by the leg.
Gewappnet Hant:	By foot over leg, by the leg, lay Messer on other leg, from bind.
Mitelhaw:	From the scabbard, Messer on the left in the *Mitelhaw.*[13]

Attacks that the technique set defends against

Schlechten Versatzung:	Strikes or thrusts from above, strikes or thrust from under.[14]
Verkerter Hant:	Strikes from above.
Gewappnet Hant:	Strikes from above, strong blows, heavy weapons, long weapons.
Mitelhaw:	Thrusts from above or under, strikes aimed towards you.

The footwork movement during the defense

Schlechten Versatzung:	Step forward or backward, leap, spring, triangle, stay in place, run in.
Verkerter Hant:	Step forward or backward, leap, spring, triangle, stay in place, run in.
Gewappnet Hant:	Step forward, run in.
Mitelhaw:	Spring out of the way,[15] step back or forward,[16] step well outward, triangle.

[13] The passages do not actually detail out where the Messer is supposed to be. They refer to the position as the *Mitelhaw*. In describing the scabbard, they give us some clue that it can start from the left hip. Either in a scabbard or where the scabbard would be. All the *Mitelhaw* plays can be completed from that position.

[14] From the *Zoll* displacement.

[15] There seems to be a stronger emphasis on the movement of springing in the *Mitelhaw*, compared to the previous techniques.

[16] The step forward seems to be an exception rather than a rule as all the other movement prefers going backwards, or well out to the side.

The movement of the Messer during the defense

Schlechten Versatzung:	Raise the arm with the point down to displace strike or thrust.
Verkerter Hant:	Shoot or raise the Messer to catch with the flat and or nail.
Gewappnet Hant:	Grab the Messer with the other hand by the point and parry.
Mitelhaw:	Hew over and down onto arm, or intercept opponent's thrust crosswise.

The *Ringen* follow-up when in *Ringen* range

Schlechten Versatzung:	Grab Messer, hand, elbow, arm wrap, neck throw, yanking with pommel.
Verkerter Hant:	Grab hand, arm wrap, striking hand,[17] yanking with pommel.
Gewappnet Hant:	Grab hand, neck throw with foot trip.
Mitelhaw:	Grab hand.[18]

In examining these passages, it is apparent that all the techniques begin from a low position. The hand typically held no higher than the girdle or the hip. Messer positions are described as at the hips, or leg, or knee. There is no mention of performing these techniques from a high position with the Messer, with hand starting at or above the shoulder.[19] It may be expected that the hand and Messer transitions to a high position to defend, or to prepare for riposte, but the starting position of the technique always described a low position.

One suggestion may be that these techniques are to be performed after a missed hew, and this can be evidenced with the idea of *wechsel* ("change"). The *wechsel* can be thought of as a position where one might find themselves in after striking downwards, and is the middle of changing to a different hew or position. Where the Augsburg texts mentions *wechsel*, their instructions go along like this: to "stop in the change" or when one "is in the change", and "raise your Messer by the leg in change". This is not the case for all the other mentions of in a low position. It tends to be more instructional, with the example vocabulary being "place yourself as such", take the Messer to X position, set your Y feet forward. This makes the starting position sound deliberate rather than incidental. It is also possible that a deliberate standardization of the starting positions helps with the learning and the

[17] Striking the opponent's hand with your hand, either upwards or downwards.

[18] The passage which describes this *ringen* technique happens with a good step forward, as opposed with a step away from the opponent.

[19] There is a passage in GF that describes starting "from day" but as a counter against the "direct displacement".

muscle memory as a practitioner need not remember much beyond that the Messer should be in a low position, or finding themselves in *wechsel*.

In examining the utility of the technique sets, it is clear that these have different roles within a whole. Firstly, the *schlechten Versatzung* covers the widest angles of attacks including thrusts. The idea seems akin to a general-purpose technique that can fit into many scenarios, and it allows for a variety of follow-ups. It works evenly with follow-ups from striking to hand grappling to body throwing distance. While the footwork is varied, the technique set seems to suggest that the practitioner should generally move around the opponent and away from his Messer.

Secondly, for the *verkerter Hant*, the passages only describe it being used in defense against strikes from above. Unlike the *schlechten Versatzung*, this application is much narrower. The footwork of the *verkerter Hant* is as varied as the *schlechten Versatzung* but with a difference in emphasis. The *verkerter Hant* passages list far more steps going forward or towards the opponent's right side, which is likely the direction of the opponent's *Oberhaw*. With more passages regarding *ringen* than the *schlechten Versatzung*, the *verkerter Hant* is likely to be thought of as an active defence to lock the opponent down after displacing. About half of the *verkerter Hant* passages involve *ringen*, as comparison to a quarter of the *schlechten Versatzung* passages involving *ringen*.

Third, the *gewappnet Hant* takes the role of weathering a powerful strike. Should an opponent use a larger weapon like a poleaxe or a two-handed sword, the *schlechten Versatzung* is unlikely to be able to directly oppose such force with only one wrist. The *gewappnet Hant* prevents the collapse of the displacement by providing two points on the Messer at both ends, the tip and the handle. The *gewappnet Hant* can be used against regular Messer strikes but very likely as an aggressive move forward to close the range to *ringen* as about 80% of the passages describe some type of *ringen* technique.

Lastly, the *Mitelhaw*. While previous technique sets mostly begin with Messer either towards the middle or on the right side, the *Mitelhaw* begins in a different position with the Messer on the left side, pointing backwards rather than forward. While described as being used to defend against almost as many attacks as the *schlechten Versatzung*, the *Mitelhaw* introduces the concept of striking the hands or arms instead of displacing the Messer. Where the *Mitelhaw* passages does describe a displacement, it tends to be against the thrust. Due to the preference for striking of the opponent's limbs, this technique's footwork involves more movement away or backwards from the opponent. This is likely to place the practitioner in the position to strike the limb from the opponent's outside rather than across the body. The passages also mention that this can be used with the Messer in the scabbard, or when the back is against a wall. As the footwork distances the practitioner further from the opponent than the other technique sets, there is only minimal mention of *ringen*. The role of the *Mitelhaw* is likely to provide an option for when the Messer is on the left side of the body. Be it in the scabbard or after a hew, the *Mitelhaw* allows for a defense to be performed from the left.

Figure 5: Messer vs sword according to Anton Rast (AR, 44r).

To sum up, while each of the technique set is distinct, they complement each other in combat by having a specific purpose. Each technique set also share the same knowledge, e.g., low positions and similar footwork, but with a changed emphasis or direction to achieve a different result. This results in a lower knowledge requirement to be able to learn all the passages. With the breakdown considered, the next paragraph will move on to considering more information as to the intended purposes of the teachings.

Looking through the passages of GF and JWM2, the words "dire position" appear multiple times in the two manuscripts. This can provide another clue to understanding the purpose of the texts. Some description of a "dire position" is as "where one can neither move backward nor forward as one wants", "against the wall", "comes at you with a sword" (Fig. 5),[20] "the opponent strikes strongly", and "from the scabbard". The manuscripts do mention that a technique can be against a strong

[20] Perhaps this can be understood as either by surprise or a larger or longer weapon.

blow, without mentioning it as in a "dire position". Thus, a "dire position" can be understood as to do with movement restriction, or unequal situation such as a larger weapon, or even being caught by surprised with one's Messer in the scabbard. It would seem that at the very least, these techniques are considered for use in situations where there are not established rules for fairness of equipment or terrain.

A passage analysis can also be done on the manuscripts. All the pieces generally follow a pattern. Longer or expository passages tend to begin with an outline of the situation:[21] the opponent is in position *A*, practitioner should go to position *B*, opponent does strike *A*, and practitioner should do defense *B* and action *B1* and *B2* till the opponent is struck or thrown. Shorter passages tend to be more concise.[22] It follows similarly to the previous, opponent does strike *A* and practitioner should do defense *B* and action *B1* and *B2*, and so on. All of the passages assume the success of the practitioner's riposte and do not mention what to do if the opponent parries the riposte. The passages are instructional rather than informative, and spend no extra space to explain much beyond telling the practitioner what to do.

With the above detailed out, one can begin to form a theory as to who might be the audience of the manuscripts. The passages are simple in their instruction, without too much beyond the initial riposte. The amount of knowledge required to perform the techniques are also kept concise, with overlaps between technique sets. The manuscripts contain knowledge about using these techniques in a number of situations where the practitioner might encounter larger weapons, or trapped, or surprised.

It would seem unlikely that the Augsburg passages were written for school fencing or for judicial dueling. While the techniques could be used for those situations, it seems more likely that the techniques were selected as a low skill barrier, straightforward but versatile system for defense. The passages do not elaborate much on the art of fencing, nor do they look at developing complex tactics.

In looking towards history for some context, one would find that there is a section of German society who was required to bear arms despite not being professional soldiers. From the 14th to 17th century, citizens in medieval towns within the Holy Roman Empire were required to take up civic duties as part of their oath of becoming a citizen of the town.[23] Such civic duties include guard or night watch duties. These duties were not necessarily daily affairs, with 17th century Nördlingen and Memmingen ordinances suggesting that one might be called up every three weeks or so.[24] With guild membership mandatory for citizenship,[25] many of these citizens would have been craftsmen such as tailors, bakers, masons, and various

---

[21] Such as the opponent is going to thrust or strike from X position. Or that the practitioner is in a "dire position".

[22] Likely because the pattern or context might have been written in a prior passage.

[23] TLUSTY 2011: 11–44.

[24] IBID: 37.

[25] IBID: 16.

others. They were not professional soldiers but had to take up a role where they might need to use their weapons in defense. It would not be a stretch to consider that the Augsburg texts were written for such people in mind. The goal of the texts seems to set forth a system for the practitioner whose main job was not fencing or fighting, nor need to devote time to learning more complex fencing skills. It had to be something that was easy to remember and still effective if and when actually needed.

3.3 Did the Technique Sets Only Exist in the Main Text?

The techniques of the Augsburg tradition of Messer fencing are not unknown or unique. The techniques were known, in varying degrees, to masters with connection to the Liechtenauer tradition, such as Paul Kal and Hans Talhoffer.[26] Kal depicts the "middle hew" in its most basic idea.[27] Talhoffer's Messer section has eight illustrations which matches techniques such as the "direct displacement", "turned hand",[28] the "middle hew", and even some of their *ringen* follow-ups. Unfortunately, neither Kal nor Talhoffer gives the techniques a clear name in their manuscripts, or it might have been easier to draw a more concrete connection.

The Brotherhood of Saint Mark (*Marxbrüder*) can also be believed to have known these techniques. Anton Rast, whose work is within AR, was a Captain of the *Marxbrüder* in 1522. Peter Falkner was another Captain of the *Marxbrüder*, first elected in 1502, and he also authored treatises containing Messer techniques. Interestingly, his Messer teachings use Lecküchner's vocabulary in contrast to Rast whose Messer section is clearly of the Augsburg tradition. However, some of Falkner's illustrations seem to have a strong resemblance in starting or in action position, though otherwise there is no mention of a direct connection. There are positions that seem to represent the "middle hew",[29] "direct displacement",[30] "turned hand",[31] and "armed hand".[32]

There are other masters including Andre Paurenfeyndt, who also teaches a few pieces with the "armed hand". Albrecht Dürer shows several positions and plays that correspond with the main texts.[33] It should be noted though, that Dürer's work (AD) seems to be a copy of other manuscripts, in particular BF.[34] The Messer section in the *Paris fight book* (P) seem to show a variation of the "direct

[26] PKM and HTM.

[27] PKM, ff 71v–72r.

[28] Talhoffer calls this the *gewenter hand* ("turned hand").

[29] PF, ff 21v, 38v.

[30] Ibid, ff 26r, 37v, 38r.

[31] Ibid, ff 28r.

[32] Ibid, ff 39r, 43v.

[33] AD, ff 73r–92r.

[34] Dürer's work also contains techniques not described in the main texts, but evidence in BF shows that he clearly had access to it.

displacement" when dealing with the thrusts, but otherwise the resemblance is superficial as one has no written text or terms to help verify.

Where Leckѝchner is concerned, there are strong similarities between some of his pieces and in the Augsburg tradition. While not direct copies of each piece, in some cases, there are identical starting positions and movements between Leckѝchner's work and the Augsburg tradition, as like the other masters. The follow-ups, however, tend to differ. Where the Augsburg tradition might suggest a hewing action for a follow-up, Leckѝchner tends to prefer a thrust action or a point-forward position. This seems to suggest that these techniques may be known to Leckѝchner, but he might have adjusted or changed the movements or follow-ups to better suit his preferences, and/or what he perceived to be better fencing.

Leckѝchner refers to the "direct displacement" as the *Pogen* ("bow"),[35] and has several follow-ups that are similar in nature, as well as some quite different. He lists the "bow" as the fourteenth *Hauptstück* ("chief piece"). The initial movements of the *Zwinger* ("constrainer")[36] have a strong resemblance to the "turned hand", and the *Zwinger* is listed as the fourth hidden hew.[37] The "armed hand"[38] is also inside his book without a name change, with various pieces and grapples (Fig. 6). Leckѝchner even has a series of techniques that resembles the "middle hew" which, he calls the *lam Stuck* ("laming pieces").[39] These "laming pieces" are techniques that aim for arms and hands joints. Two techniques within the "laming pieces" begin as one would with the "middle hew".

As one can see, the pieces of the Augsburg tradition and Leckѝchner do overlap in terms of starting position and movement. However, the techniques themselves are not identical, and often have modified follow-ups or a different emphasis. It is clear that the pieces and techniques from the Augsburg tradition are not unique; their core movements and ideas have been depicted by masters from different traditions including Leckѝchner.

What makes the main texts interesting is the choice of the technique sets: the "direct displacement", "turned hand", "armed hand", and the "middle hew". The Augsburg tradition choses these techniques to focus on, while they do not seem to be the main focus of other masters. This generalization excludes Talhoffer, whose illustrations in his Messer section would otherwise seem to correspond well with the text of the Augsburg tradition. It would not be inappropriate to say that Talhoffer's Messer teachings fall within the Augsburg tradition, except for the fact that he did not call them by the same names or describe their movements.

---

[35] LM, ff 23r, 42v, 44r, 44v, 128v, 129r.

[36] *Zwinger* is not an easy word to translate to English. In German, it refers to part of a castle that constrains or narrows an enemy force between two walls. It has connotations to oppressing or pressing. Sometimes translated to fortress, prison, narrow space, cage. At other times, it is left untranslated. FORGENG 2015 translates it as "constrainer". For further discussion of terminology translation in Leckѝchner, see FINLEY 2021.

[37] LM, f 28v.

[38] IBID, ff 138v, 139r, 140r.

[39] IBID, ff 184v, 186r.

Figure 6: The armed hand according to Hans Lecküchner (LM, f 143v).

One finds that the masters who use Lecküchner's vocabulary tend to place emphasis on drastically different type of techniques, such as the six hidden strokes[40] and the 17 *Hauptstücke*.[41] These are far more encyclopedic and complicated than that of the Augsburg tradition. In contrast, the Augsburg tradition focuses on four types of techniques, with a large portion as variations. It essentially uses a smaller skill set to tackle common attacks.

Another interesting observation to make is the possible inference of context through some of the techniques taught. The Augsburg tradition contains paragraphs about what to do in a dire position, a situation in which there could be a numerical, positional, or equipment disadvantage. It seems to account for

[40] *Zornhaw, Wecker, Entrusthaw, Zwinger, Winker, Geferhau.*

[41] Four Positions, *Versecz, Nachreisen, Uberlaufen, Absetzen, Wechsel Durch, Zuck, Lauff Durch, Abschnitt, Druck, Ablauffen, Benehmen, Durchgehen, Pogen, Messer Nehmen, Hengen,* and *Winden.*

situations where the opponent might have bigger weapons, such as a two-handed sword or spear, or when the Messer is in the scabbard as the opponent strikes. This directs the idea that perhaps the teachings of the Augsburg tradition have an eye towards situations not typically encountered in dueling or school fencing. Self-defense with a Messer could also be a possible factor in leading to the type and number of techniques chosen to represent the Augsburg tradition.

While contemporary manuscripts such as Munich 582 has spent much ink with passages in a similar pattern to those of the Augsburg tradition, Leckürchner's work also has much that goes beyond detailing out only one attack and one response. Leckürchner also elaborates about riposte, counter riposte, and technique possibilities. As well, he illuminates the dynamics of Messer combat through providing a conceptual framework for fencing. Leckürchner details the art of *Messerfechten* to a higher degree of complexity for his readers and explains techniques together with ideas such as *vor* ("before"), *nach* ("after"), *indes*,[42] *stark* ("strong"), *schwach* ("weak"). With a conceptual framework in place, Leckürchner can account for multiple shifts in combat dynamic during a single engagement. These passages help to illustrate the fluidity of combat and how to take and/or retain the upper hand.

While dynamic shifts in combat are expected, it is likely not the main focus of the Augsburg tradition authors, as such concerns falls out the purpose of their manuscript, which is essential self-defense with the Messer vs learning the art of Messer fencing.

4 Conclusion

The study of medieval German Messer fighting revolves around Leckürchner for two reasons: the connection between Leckürchner and Liechtenauer helps many students of the latter to approach the text of the former, and the similar[43] language and concepts provide a comforting bridge. Leckürchner's work is also the most encyclopedic manual on Messer fighting in current known existence. It covers many techniques as well as fencing concepts. These reasons help to explain why discussion around Messer fighting always tends to converge at Leckürchner.

That being said, diving straight into Leckürchner without prior Messer instruction can lead to a very slow study even for experienced fencers. It can be nearly impossible for those who have little fencing experience. An alternative for beginners could be to work through a shorter series of techniques which help to illustrate Messer handling. Studying the Augsburg tradition can provide such an alternative to jumping straight into Leckürchner.

There are several benefits to this decision. The first is that the Augsburg tradition uses a tighter scope of techniques, which are less daunting to tackle. The second is

[42] *Indes* is a concept that is difficult to translate into one word. Some translations include "at once", "immediately", "instantly".

[43] Although terms and concepts between Leckürchner and Liechtenauer are not always the same.

that learning another style of Messer fighting can complement that of Lecküchner's teachings. The third reason is that when one does pick up Lecküchner, there would already be a good Messer handling foundation. Approaching Lecküchner and working through his work with the ability to physically maneuverer through Messer fighting is an advantage to one's study. This will allow one the best position to enjoy Lecküchner's teachings.

The Augsburg tradition is different, but it is not unique. The techniques are also known to many masters connected to Liechtenauer and Lecküchner. Even Lecküchner displays some of these techniques in different forms. What makes the Augsburg tradition distinct is the deliberate focus of the four technique sets. The selection is compact, but comprehensive. This means that the different sets of techniques and their variants are sufficient enough to account for basic Messer combat. While it is not as thorough in covering every possibility and eventuality as Lecküchner, it does have a good set of responses for different attacks, ranges, and situations. The grouping of techniques, and the emphasis of focusing on these four sets, seem a deliberate choice. Moreover, it seems to be a deliberate choice that persisted from manuscript to manuscript, alongside Lecküchner's teachings.

APPENDIX: TRANSLATIONS

The intent of these translations is to provide a common style between the main texts, so I also include BF even though its techniques are translated earlier by CHRISTIAN HENRY TOBLER.[44]

I have translated them with the purpose of flowing well in the English language, as this makes readability easier when comparing passages across manuscripts.

Each passage will also be marked to show which technique set it belong to in parentheses. I have chosen to keep the common terms in German, together with its translation. I believe this helps when the reader decides to compare different translations.

As with all translations, the lens of which the translator understands the text will always affect the translation. I have strived to keep true to the text and allow the intent of the author to shine through. If there are mistakes, please pardon me.

BAUMAN'S FIGHT BOOK (BF), CA. 1470S, SOUTHERN GERMANY

Item (VH): When the opponent fights you with a Messer and hews at you from above. Parry with your Messer on the outside with the flat, and on the nail. Go forward swiftly with the left foot and fall in with the pommel over his arms, as it stands illustrated, and hew him in the head.

Item (VH): Note, a good play. When the opponent hews in down to you from above. Catch the strike on the nail, as before, and go with the left foot immediately. Hit his right downwards towards the ground with your left hand, and hew in onto his head, as is illustrated here.

Item (SV): Note, a good play. When the opponent strikes at your head from above. Wind in with your Messer from below and catch the strike thereon. Fall in with the pommel over his arm and yank him to you, stepping to with the left foot. Drive with the left hand around his neck, as it stands illustrated, so throw him on his back.

Item (SV): Note, a good play. When the opponent hews in down from above. Wind through below with the Messer and run in. Drive with your left hand high above his right arm and grasp beneath the shoulder and stab through with your Messer yourself and his neck, as it stands illustrated.

Item (VH): The opponent hews from high to your head. Parry as before with the flat and on the nail and likewise, send his Messer away. The opponent must strike again, and when he will strike, so seek the nearest and pull off his arms, as it stands illustrated.

Item (GH): Another play, for a strong one. When the opponent hews a strong blow at you. Grab the Messer with *gewapnoter hant* ("armed hand") and catch the strike thereon. Step to him with your left foot and lay the Messer on his throat, as it stands illustrated.

[44] For CHRISTIAN HENRY TOBLER's translation, see pp 38–41 in this volume.

Item (GH): When the opponent hews in from above to your face. Hold the *gewapnoter hant* ("armed hand") and catch the strike on the Messer. Wind to the head over his arm and yank backwards, as it stands illustrated, so break his arm or take the Messer.

Item (GH): When the opponent hews in from above to the head. Hold the *gewapnoter hant* ("armed hand") and catch the strike on the blade, as before. Step hard behind him with your right foot and lie on his head or neck, as it stands illustrated, and throw him onto his back.

Glasgow Fight Book (GF), ca. 1508, HRE

Take note: Some plays with the Messer.

Item (SV): When the opponent strikes or thrust from above. Place yourself as such. Set the left foot forward and hold your Messer in your hand beside the right hip. When the opponent hews to your head from above so step to him with the *schlechten versatzung* ("direct displacement") such that the edge sits above. Leap with the right foot to his left side and hew to his head or where you can reach so that he is yours.

A Second Play

Item (SV): When the opponent strikes at you *von dach* ("from day"). Set the right foot forward and parry, springing towards his right side in the *drÿ angel* ("triangle"). Strike or stab him where you will. When you spring, raise the other foot a little so that you may better pass him with the leap.

Take note: A position as needed

Item (VH): Set the right foot forward and stop your Messer in *wechsel* ("change"), that the sharp point is downwards. When he strikes *von dach* ("from day") so shoot the Messer up and parry with the *verkerter handt* ("turned hand"). Leap towards his right side in the *dryangel* ("triangle"). Hew to his head or stab at him.

But another position

Item (SV): When the opponent fights long at you from under. Stand with the left foot forward and raise your Messer by the leg in *wechsl* ("change"). When he attacks you, take off with the Messer where the hilt stands above and the point below. Parry so, (and this particular parry is named *zoll versatzung*), and spring to his left side in the *dryangel* ("triangle") and hew or stab.

This is for when one is in a dire position.

Take note (GH): When the opponent strikes strongly at you *von dach* ("from day"). Go with the *gewappnet* "armed hand" and parry resolutely. Wind the point in against his face and slice in with the edge, down at the hand.

Again, a good play or position.

Item (MH): When the opponent thrusts at you from below. Place the right foot forward and bring your Messer into the *mittl haw* ("middle hew"). Hew in athwart his Messer and weaken his thrust. Wind your Messer on his Messer and step with the left foot forward. Allow the Messer to turn towards the head, and hew or stab.

Again, a play for the heavy weapon.

Item (GH): When the opponent strikes at you *von dach* ("from day") with a heavy weapon. Parry with the *verwappenter handt* ("armed hand") and wind in your point against his face. Stab into his chest or, step in with the left foot behind his right foot well in the *wag* ("balance") and seize his neck with the Messer and throw him.

Again, a play and a slice.

Item (SV): When the opponent fights you from below. Set your left foot forward and hold your Messer above your right leg in the *wechsl* ("change"). When the opponent stabs or strikes, step at the same time and with the tip pointing downwards, parry with the half parry. Grab him by the Messer with the left arm, and let the Messer proceed straight up with the edge going through the mouth.

A Messer taking.

Item (SV): When you strike at the opponent *von dach* ("from day") and he parries you with the *schlechten versatzung* ("direct displacement"). Grasp him on the Messer with your left hand, such that the thumb is against his hand which has the Messer. Take the Messer out of his hands and throw it away.

Again, another grabbing.

Item (VH): When the opponent hews at you *von dach* ("from day"). Go up with the *verkerter handt* ("turned hand") and parry. With the parry, wind the hilt over his arm and grab him by the elbow. Take it upwards, and pull the arm downwards so from that he falls on his mouth. The Messer must be allowed to go well around the arm.

A play from the scabbard, and when in a dire position. As well as where one can neither move backward nor forward as one wants.

Item (MH): Take note of when the opponent stabs or strikes to you. Take the Messer in the *mittlehew* ("middle hew") with your right foot forward. When he proceeds to you, hew from your side *von dach* ("from day"). Strike above and over, and spring out of the way from him out towards his right side in the *dry angl* ("triangle"), and hew him onto his right arm.

Anton Rast (AR), ca. 1553, Augsburg

Item (VH): When the opponent goes after your head with a Messer. Defend with your thumb on the flat of your Messer and catch the strike on the nail, as it stands illustrated. So that you are free to work what you will as you will find here after ~

Item (VH): So, you have parried as positioned before. Step forward with your left foot and fall in over his arm with your Messer. Hew in from above onto the head, as it stands illustrated, that you might also cause him to fall over with the hit, and have his head go onto the ground.

Item (VH): When the opponent strikes over to the head. Parry with the flat, as before, and step forward with the left foot. Grab forth with your left hand onto his right, and thrust him downwards to the earth. Hew in through the head as it stands illustrated.

Item (SV): When the opponent goes after your head. Wind in through with your Messer and extend yourself long inward. Fall in, with your left arm inverted, around his arm and ſtrike in under your armpits. Stab with your Messer between yourself and his throat as it ſtands illuſtrated.

Item (GH): When the opponent hews from above, after the head. Hold your Messer in *gewappneter handt* ("armed hand") and run in under the ſtrike. Wind in with the point at his neck and ſtep behind him with your left foot, as it ſtands illuſtrated. Throw him onto the back.

Item (Others): When the opponent runs to you swiftly. Do so as well. After the hew to the head, turn yourself around in front of him and hew him long after his arm as it ſtands illuſtrated.

Jobst von Württemberg (JWM2), ca. 1556, Augsburg

Here are the fencing plays in Messer.

Item (VH): Stand with the left foot forward and hold the Messer by the right foot over the leg. Drive up with *verkertter hand* ("turned hand") and spring in, towards his right side. Hew in long to him, after the arm or towards the head or where you will.

Item (VH): Stand as before. Rise with the parry, springing off towards his left side. Hew in long to him, to the arm or neck or where you will.

Item (SV): Stand as before. Rise with the *schlechten versatʒung* ("direct displacement") and ſtay in place. When his hew is upon, so grab him on the inside of his hand, then lift the hand which has the Messer and hew or ſtab.

Item (SV): Stand again as before. Rise again as before, and grab him at his elbow and lift it upwards. Push him from you and hew or ſtab.

Item (SV): Stand again as before. When the opponent hews at you so proceed as before, and drive your left arm around his right arm. Lock his right arm by setting your left fiſt upon your cheſt so that he would not be able to *ringen* ("wreſtle") and hew or ſtab.

Item (SV): Stand again as before. When the opponent hews crosswise to your left side, so let the point hang. Grab him on the inside of his hand that has the Messer and wrench his hand, and hew or ſtab.

Item (GH): Stand again as before. When the opponent ſtrikes or hew at you, so rise with the *gewapfentter handt* ("armed hand"). Wind the point in to his face and ſtab in.

Item (MH): Stand again as before with the right foot forward and take the Messer into the *mitelhaw* ("middle hew"). When the opponent ſtands towards you with the right leg forward, so pull your right leg with a ſtep backwards and hew crosswise from the *mittlehaw* ("middle hew") to the opponent's arm.

Item (MH): When you are in a dire position. Take your Messer again in the *mittelhaw* ("middle hew"). When the opponent thruſt at you, hew crosswise well from the *mütelhaw* ("middle hew") and spring out to his right side and hew.

Item (SV): When you are in a dire position as before. Place the left foot forward and take the Messer onto the right leg. Hang the point down with the *schlechten versatzung* ("direct displacement"), springing to his left side and hew or stab, rising up as he strikes onto you.

Many good plays with the Messer.

Item (VH): Stand with the left foot forward. Remain in place and parry with the *verkertter handt* ("turned hand"). Hew around the head to the right arm.

Item (VH): Stand as before. Step with the right foot forward and then backward, parrying with *verkerter handt* ("turned hand"). Hew around the head to the right arm.

Item (VH): Stand with the left foot forward. Step forward and parry with the *verkertter handt* "turned hand". Spring to the back of the opponent towards your right side, hewing him to the head.

Item (SV): Stand again with the left foot forward. Use the *schlechte versatzung* "direct displacement" and spring off to his left side. Strike the opponent to his head.

Item (VH): Stand with the left foot forward. Parry with the *verkerter handt* ("turned hand"), stepping with the right foot forward. Hit his right hand with your left hand, upwards or towards the back and strike to his head.

Item (SV): When the opponent thrusts at you. Stand with the left foot forward. Step with the right foot forward, pushing his thrust, in the manner where your point is facing down, and hew in onto his left side, towards the head and to the neck.

Item (SV): When you are in the *wexel hawst* ("changing hew") and have your Messer pommel by the belly and the point in front of you. When the opponent strikes to you, so stay in place and parry with the edge and hew him to the head on his left side.

Item (SV): When he strikes to you and you stand as before. Step forward and parry with the edge, and step once more backwards and strike him to the head.

Item (MH): Set again in the *Mitelhaw* ("middle hew"). Stand with the right foot forward, and when the opponent stabs down at you from above, so push his thrust away with the back of your Messer and spring off towards his right side. Hew into the head.

Item (Others): Stand with the left foot forward and lay your Messer by the right leg. When you both hew, fall into him with the right hand and press his Messer to him with your left hand over his right hand with the edge on his neck.

Item (MH): Stand with the left foot forward and position yourself in the *mitel-haw hawt* ("middle hew"). When he come at you, tread with a good step with the left foot forward and drive up with the edge. Parry, and grab with the left hand on the right arm and raise the Messer and stab or strike.

Item (Others): When the opponent wants to fall in onto you from above. Pull your right foot backwards and turn around the hand that has the Messer so that you have struck his hew and stepped backwards.

Item (Others): When you both have struck again, bind and fall in with the Messer. Go over his arm and lay the Messer over your leg against your groin, and step with your left foot behind your opponent. Grab from behind through the legs, lift him up and throw.

Item (Others): Stand with the left foot forward. Strike upwards to both side against the man. Step with the right foot forward and hew around the head with the *mitelhaw hawt* ("middle hew"). Strike up against the man and parry with the edge, and hew over the head once more in the *mitelhaw hawt* ("middle hew").

Item (Multiple[45]): Stand with the left foot forward. When there are two who hew to you onto the head. With one of them standing behind you, who will hew the same hew second. Parry the front and step with the right foot forward. Hew backwards, and parry the one behind. Always turn yourself standing with your left foot forward.

Item (GH): Stand with the left foot forward and have your Messer lying by your right leg. When the opponent strikes to you with a poleaxe or with any long weapon to your head. Step with the right foot forward and grab the Messer with the other hand by the point. Parry high with the edge over your head and grab his right hand with your left hand, and hew into the head.

Item (GH): If you have both struck, so fall in with *gewapfneter handt* ("armed hand") above his right arm, and stab into the throat or onto the chest.

Item (MH): Stand with the right foot forward. Take your Messer in the *mitelhaw* ("middle hew") by the left side. When he strikes onto you so pull the right foot backwards and hew into the hand. With the strike, allow the Messer to go around the head so that you lift him in the hand.

Item (MH): When he stabs at you similar to before. Stand again like before with the right leg forward and lay yourself above in the *mittelhaw* ("middle hew"). When he stabs for your thighs, step backwards and hew in crosswise after his hand and pull firmly with the Messer. Hew him to his hand towards his wrist.

Item (LV): If you want to take a Messer with one hand. Defend with the *langen versatzung* ("long displacement"). When he binds you, run in over his right hand and arm. From the side, go above and down, and throw the right arm upwards. Yank his Messer out with one hand towards his body or where it will cause him grief.

Item (LV): When the opponent strikes at you with *langen versatzung* ("long displacement"). Bind with him, and run in over the outside. Take the elbow down and away. Step in with the left, wrestle and throw him over your left hip.

Item (LV): When the opponent strikes to you with the *langen versatzung* ("long defense"). Bind with and run once more to the outside, over and above. Take hold with the hand, the Messer in front and throw onto your left side over the shoulder, over and over.

Item (LV): When he strikes at you as before. Bind with, and run in again from outside over the opponent. Set the left forward, meeting the back of his right foot and grab him inside where it is soft. Throw him over the thigh so that he falls.

---

[45] Compatible with either "direct displacement" or "turned hand", as both displacements describe stepping forward and then back with a strike.

Item (LV): When he strikes at you as before. Bind with and run in again to him, from above on the outside. Step to him and press the Messer on the outside of the neck, and choke hard. Grab him by the right leg and raise him up high and throw him. In order to slice him well, you must really press forward with the Messer at his throat. And if you wish that he be thrown, press the crossguard in hard at the throat.

Item (VH): Stand with the right foot forward. When he strikes to you, you ought to have your Messer lying by your right leg. Go up with the point and reverse the Messer. Step to him and grab him on the inside of the arm with your left hand. Lock his arm under your arm and in front of him so have him taken and caught.

Item (VH): Stand with your right foot forward. Step in again as before and hold him by the hand, on the inside with your left hand on his right hand that has his Messer, and bring him down.

Item (MH): When in a dire position with the Messer. When the opponent has you against the wall. Position the high right foot forward and lay the Messer in the *mitelhaw* ("middle hew") onto the left side. When the opponent strikes to you and from the back, so step with the left foot going outwards and the right foot to the back after as well. Hew over and down into the arm or to the hand.

Item (MH): When he stabs at you while you are in a dire position. Stand again as before, and step well going outwards again. Hew from over, downwards into the hands or to the arms and, so that you cause his Messer to fall and thus you have won.

Item (GH): When he comes at you with a sword and have you in a dire position. Stand with the right foot back and lay the Messer on the other leg, and when he strikes in downward from above so grab the Messer by the point and go up *gewap-ffneter* ("armed hand") and yank in the forward hand so that you have him.

Here is the end of the Messer fencing.

Item das vierd twirch treib also wen du mit aim
ringst in den armen so prich im auf mit dem
rechten arm zu ge ein weil und mit im so verbin
so spring mit dem rechten fuss hinder seinen rechten
fuss und mit dem rechten elpogen hinder sein rechte
uchsen als hie gemalt stet so wurffstu in auf
den rucken das get zu paiden seiten und must
dich auch woll fursichsten das er dich nit
zu ruck stoss und das tu in in die wag pringst

«DAS RINGEN WILL HABEN DREYERLAY, STERCK, MASß, VND PHENTIKAIT»: COMPARING THE RINGEN OF BAUMAN'S FIGHT BOOK TO MODERN GRAPPLING STYLES

Elias Flatscher
University of Zurich

Wrestling requires proper posture, coordination, technical-tactical skills and being able to read the opponent's intentions,[1] that is as true today as it was roughly 550 years ago. The newer section of *Bauman's Fight Book* (BF) is probably the most extensive surviving source on grappling and unarmored fighting in medieval Germany, devoting 95 out of 150 pages (63.3 %) to these topics, making it a quite unique and highly influential source. However, the author does not tell us how to develop these qualities, nor how to ingrain the techniques until they become second nature. Why put so much emphasis on some aspects and completely ignore others? How did one become a feared grappler in the 15th century? And how do modern grappling styles such as wrestling and Judo compare to the medieval *Ringen* in BF?

1 "CLEAN" AND "DIRTY" WRESTLING, THEN AND NOW

Ringen—in the sense of stand-up grappling—was without any doubt a key discipline in medieval European martial arts. In fact, a case could be made that in the 15th century, it might well have been *the* key discipline and the discipline that should be mastered first, since many of the movement principles, mechanics and techniques in armed fighting could be described as "wrestling with and for the weapon"[2]. This argument is supported by the famous quote from the *Nicholas Pol House Book* (N), "And know that all martial skills[3] come from wrestling, and all of fencing comes basically and thoroughly from wrestling".[4]

There might be other reasons to learn wrestling first: learning any grappling discipline, but especially stand-up grappling and throwing will take time. It is commonly believed that developing a high level in stand-up grappling requires ten

[1] Analogous translation of the title, a quote from BF, f 15r, based on the description of those qualities in the text.

[2] WELLE 1993: 3–4.

[3] Lit. "courtliness".

[4] N, f 86r: "*Vnd wisse das alle höbischeit kompt von deme ringen vnd alle fechten komen ursachlich vnd gruntlich vom ringen*". Part of this quote was used by RAINER WELLE as title for his dissertation (WELLE 1993), which is still probably the key overview on medieval German wrestling treatises (expanded by his later publications).

years (or the equivalent of roughly 8.000–10.000 hours[5]) of dedicated training under competent instruction, preferably starting at age 12 at the latest and having developed good coordination first; a passable level may be reached in 1.000–2.000 hours. Seeing that wrestling was likely about as common a pastime in medieval Europe as soccer is nowadays, it can be assumed that most adolescents interested in making fighting their career probably had racked up at least 1.000 hours of wrestling practice by the time they started training seriously with weapons, and then possibly expanded on it by adding armored wrestling. Also, other common medieval pastimes[6]—and at least in some cases manual labor—were arguably beneficial in developing physical traits beneficial to both wrestling and fencing.

Wrestling with and without weapons is therefore prominently featured in most 15th century fencing manuscripts. Like most 15th century sources, BF focuses on combat wrestling (*Kampfringen*), not sportive wrestling ("*gesellen ringen*"). As the "Not more than a foreword"[7] states, a lighter wrestler is at a distinct if not unsurmountable disadvantage in a sportive context; however, the implicit intention of the author seems to be to prepare the reader to be able to hang in there with a stronger and heavier person "*im ernsten*" (meaning combat wrestling) by teaching him "fighting techniques and killing techniques".[8] However, other sources imply a strong connection between sportive and combat wrestling: Pietro Monte's comments on medieval wrestling rulesets all over Europe are based on the argument that some of them enforce—or prevent—habits that would in his opinion be beneficial or detrimental in armed combat.[9] While ground fighting was historically speaking a rarity on battlefields around the world, certain scenarios that the German treatises often were written for, such as judicial duels (requiring the downed opponent to be finished off with the dagger)[10] or taking and holding prisoners[11] make ground control an essential element of fighting.[12]

---

[5] Top grapplers usually are on the mat at least five days per week, often twice per day, typically completing 11–16 training sessions (16–40 hours) total per week (including conditioning).

[6] Talhoffer recommends throwing stones, drills with staffs, jumping, dancing, drills on horseback (see HTK, f 1v), etc. Monte praises wrestling above all other forms of training and additionally recommends throwing various implements, acrobatics, running and vaulting on horseback (see FORGENG 2018, esp. 147, 376–400). Johannes Rothe's *Ritterspiegel* (1410) adds strategy games, swimming, archery, as well as climbing ladders, poles and ropes to the list (see HUBER & KALNING 2009: 184–185).

[7] *Nit mer dan ein vorred* (BF, f 15v).

[8] *Kampf stuck vnd mord stuck* (BF, f 15v).

[9] See FORGENG 2018: esp. 371–376.

[10] BF, f 103r.

[11] BF, f 73r–74r.

[12] Despite being strongly opposed to including ground fighting in regular wrestling, Monte also gives recommendations on how to continue wrestling on the ground in duels; see FORGENG 2018: 219–200.

In fact, most cultures around the world seem to have used some form of sportive stand-up wrestling as an essential part of their preparation for war and the base for hand-to-hand combat,[13] and the techniques and rulesets of modern grappling sports have evolved from these traditions.[14] Only in the last century or so, wrestling in the military has been more and more often supplemented or even supplanted by boxing and/or ground grappling. It is therefore hardly surprising that very often, experienced wrestlers such as GEORGE DE RELWYSKOW or IMI LICHTENFELD created hand-to-hand combat systems for military and law enforcement.

The newer section of BF also features an unusually large number of "dirty" techniques, many of which would only be effective unarmored, including punching—possibly making it the handbook of choice for the medieval pub brawler,[15] similar to Pascha or Petter in the 17th century. Some of these techniques can be found in modern MMA, others—surprisingly—in grappling sports: while they aren't typically described in instructional media of the sport,[16] all my coaches taught illegal techniques to their athletes (but never to people not competing for the team) and explained how they are successfully hidden in plain sight during competition. These include messing up finger joints, levering elbows, shoulders and knees, head butting, "oil checking"[17] and "rips",[18] and use of equipment.[19] Basically the only fouls that aren't taught are obvious ones like punches and kicks higher than the knees—lower kicks can be concealed as sweeps. These explanations serve a dual purpose: first and foremost, they should heighten the wrestler's awareness of illegal techniques performed by his opponents; secondly, if he so chooses, he can implement the illegal techniques himself—which is sometimes referred to as "dirty" or "hard" wrestling or "dark side of the moon", but terming them "hidden techniques"[20] as on folio 38r in BF might be equally fitting—and by hiding them, not

---

[13] See COSNECK 1944.

[14] In turn, WELLE argues that sportive wrestling in medieval times evolved from combat wrestling. See WELLE 1993: 352–353.

[15] A brawl resulting from an everyday civilian situation is clearly hinted in f 70r, which starts with a verbal altercation.

[16] An exception being KESLER 2016.

[17] A medieval example of an "oil check" in a duel comes from the description of the siege of Rovereto in 1487: Johannes, count of Sonnenburg, fights a duel with a Venetian Noble. They end up wrestling and falling to the ground, Johannes on the bottom. He however forces his left hand in the opponents unarmored rectum, who starts "moaning" loudly and calling on Saint Catherine, at which point the four seconds intervene (see WELLE 1993: 224–225).

[18] A "rip" is essentially a joint lock that isn't pulled in a controlled manner, but "ripped" through, resulting in immediate injury rather than giving a chance to submit.

[19] I was instructed to shop for a Ranggeln belt fitting the following criteria: thick leather and a sharp but inconspicuous brass buckle to rub across the faces of my opponents in opportune situations.

[20] *Verporgen stuck.*

be caught in the act. Even if a foul is called, a wrestler is usually not disqualified from a match the first time around if the foul is deemed unintentional, and if the opponent is injured, victory may be given to the offending wrestler even if the injury happened through a foul (although, statistically speaking, most injuries are the result of not wanting to give up points). In most cases, fouls aren't committed to injure or cripple an opponent, but in order to off-balance him as a reaction to the surprise, pain, and frustration.

2 Discussion of the Tactical Advice of *Bauman's Fight Book*

The general advice on wrestling tactics of BF is mostly contained in the introduction.[21] It shows a clear influence of the *Ringen* school of Ott Jud, but goes beyond the usual extent in terms of interpretation and comments[22] and addresses how one should wrestle with weaker, equal and stronger opponents. Tactical advice is rarely featured modern books on wrestling and Judo, instead we usually find information on biomechanics, off-balancing a person, and phases of a throw, as well as fundamental positions, how to move on the mat, and sometimes also training exercises and drills.[23] These are not mentioned in BF; while it is possible that pages from the wrestling section are missing, it is also possible that—being a manuscript—it was written/compiled for a very specific audience, maybe even a single person, and intended to be used within a living line, in which the corresponding information was most likely transmitted orally.

The verbal instruction grapplers get today in regard to lighter and heavier opponents/partners is remarkably similar to the one in BF. When training with partners who weighed 100–120 kg (compared to my 74–84 kg), I was instructed not to "hang together" (avoid "*Ringen aus den Armen*" or "*Gleichfassen aus den Armen*") but rather break all gripping attempts, either wrestle without tie-up ("*Ringen aus dem Zulauf*") or securing an outside 2-on-1, keep the opponent moving, to retreat if necessary (never in a straight line but circling to avoid constant pressure), and to preferably attack with outside single legs, arm drags, slide-bys, shoulder throws, various hooks and trips (see Table 1 for similar techniques in BF). When competing against lighter or less skilled opponents, the advice is usually to end the match quickly—ideally taking them down immediately and never letting them get back up—since a lot of things can happen as the match drags out. A teammate of mine (ranked 5th in Dagestan) once failed to stick to that unwritten rule, resulting in an open skull fracture (from an "accidental" headbutt) which ended not only the match but also his wrestling career.

[21] BF, f 15rv

[22] See Welle 1993: 124–128.

[23] See for example Inokuma & Sato 1979: 11–22; Mysnyk et al. 1994: 1–4; Rothert & Tepper 1988: esp. 10–23 (additional drills for every technique are presented throughout the book); Rothert 1994: 3–44. For a more in-depth study of biomechanics in wrestling and Judo, see Petrov 1986: 152–165 and Scott 2019.

If the opponent is equal in strength/size, the BF suggests proceeding cautiously and preventing openings which would allow the opponent a clear positional advantage, making takedowns easy. To this end, maintaining the proper balanced stance ("*wag*") in motion and avoiding off-balancing attempts is paramount. Of particular interest is the sentence "and always hang[24] on him heavily and wear him out[25]".[26] This instruction is often repeated by modern wrestling coaches. My coach JOCHEN ENGELHARDT even still uses the same wording in his instruction ("*abar-beiten*"). Apart from making the opponent carry one's weight as much as possible (without getting out of position and opening up to a counter), "wearing him out" is typically achieved through making him react to attacks, both fakes and real attacks, since according to my coaches, a counter will take twice as much energy as a proper attack.

3 DISCUSSION OF THE TECHNIQUES OF *BAUMAN'S FIGHT BOOK*

3.1 COMPARABLE TECHNIQUES IN MODERN GRAPPLING DISCIPLINES

Any attempt to draw comparisons between medieval *Ringen* techniques and techniques in modern grappling disciplines is necessarily an interpretation of the source material, and the human mind is tempted to associate what it sees with known patterns. Therefore, the techniques listed here shouldn't be seen as "equivalents", but rather as a base for further study. That being said, knowledge of modern techniques can help to reconstruct medieval techniques, especially if the biomechanics aren't overly accurately described and/or depicted in the treatise. For example, BF often states, "turn yourself around before him",[27] without mentioning how, while in modern wrestling and Judo, at least four ways of turning for a throw are taught.[28]

In Tables 1–7, I have included analogous techniques from Amateur wrestling (Freestyle and Folkstyle), Catch wrestling, Judo and MMA. An expanded version—systematically including medieval/early modern authors and/or modern grappling disciplines may be attempted in the future. Pages where the technique is being countered are put in square brackets;[29] pages where it is mentioned as an option are put in round brackets; however, they are still included in the total number of times that technique is featured.

[24] Lit. "sit", implying a dropping of one's weight.
[25] Lit. "work him down".
[26] BF, f 15v: "*vnd alzeit swärlich in in seczt vnd in ab arbaiczt*"
[27] "*ker dich vor im umb*"
[28] See ROTHERT 1994: 35. In addition to the three variations pictured there, a jumping entry (turning 180–270°) is also possible.
[29] Folia where the technique is first described and then countered are not written in brackets.

Table 1: Amateur wrestling: takedowns and takedown counters.[30]

Technique	Folia in BF	Total
Outside hook/outside leg or knee block/sag throw	16v, 17v, [18r], 33v, 34v, 41v, 42v, 45r, 47r, (47v), 52r, [52v], 54r, 59r, 61r, 64v, 65r, 70r	18
Outside trip/step across trip	18v, [19r], 19v, [20r], 20v, [33r], 69r, 72r	8
Outside single leg	33v, [34r], 44v, (47v), 50r, [50v], 52v, 66v	8
Inside single leg	16r, [16v], [17r], 46r	4
Inside single leg with back to opponent	58r, [58v]	2
Spear double leg	42r	1
Inside hook/Grapevine[31]	(17r), (46v), [47r], [47v], 48r, [48v], [49r], [49v], (51r)	9
Hip toss/hip throw	(17r), 43v, (46v), (48r), 49r, (51r)	6
Outside hip throw	36r, 62v	2
Sidakov/Saitiev slide-by[32]	71v	1
Shoulder throw	39v, [40r]	2
Outside shoulder throw	38v, 39r, 60r, [60v]	4
Fireman's carry	36v, [37r], 49v, [50r], 50v[33]	5
Bodylock lift	[43r], [51r], [51v], [53r], [53v], [54r], [61r], [61v]	8
Rear bodylock lift	[57r], 57v, [58r], [59r], [59v], [62r]	6
Knee tap	35v, [36r]	2
Falling arm throw	18v (variation: grip on neck), 35r, 43r (reverse)	3
Rear crotch lift	[40r]	1
Arm drag	20r	1
Sprawl	16v (standing variation), 37r, [37v], [38r], [38v]	5
Roll	37v, 38r	2

[30] Terminology based on PETROV 1986 and MYSNYK ET AL. 1994.

[31] The inside hook/grapevine is technically speaking a position rather than a takedown. From an inside hook, at least three throws are possible (Judo terms in brackets): a hip throw with a hook (*Uchimata*), an inside trip (*O Uchi Gari*) and a grapevine throw or "*Ossi-Schleuder*" in Nuremberg (*Kawazu Gake*), which can range anywhere from a foot sweep to a rolling throw (similar to *Sumi Gaeshi*) all the way to a full suplex—which is fairly unlikely here, since there are no other suplex variations included. It is not clear which throws are intended here.

[32] RUCH ET AL. 2014: 131. Also see section 3.3 for discussion.

[33] The technique on ff 49v/50v is difficult to classify: it could be called a headlock hip throw or an unusual fireman's carry variation.

In wrestling, we can find analogous techniques for most of the throws and takedowns (Table 1). We can see a clear overall preference for outside hook variations[34] (18) followed by body-lock lifts (always countered, but with some successful re-counters), followed by single leg variations (14 each), inside hooks (9), hip throws (8), outside trips (8) and various shoulder throws (6). It should be noted that all of the shoulder throw variations depicted are illegal in wrestling since only the forearm is gripped, usually with both hands.

While almost all the takedowns in the section fit into modern wrestling terminology, they aren't all common versions. For example, the outside hooks are usually done either with the front knee on the mat, bending the front leg with the knee far over the toes or trapping the opponent's leg between the pinched thighs in order to prevent inside hooks. The outside single is usually done with both hands on the opponent's leg while pressing/ramming the side of the head—rather than a hand—into his chest; still, the principle remains the same. The reason for this is that the stance is lower, therefore breaking the balance backwards takes more force; also, the modern variation allows for a larger variety of finishes, including lifts. The variations depicted would be called "snatch single with shoulder hit" in wrestling,[35] are only used against upright opponents and therefore more commonly used in in Judo (*Kuchiki Taoshi*, illegal since 2011/13) and Sambo.[36]

The arm drag (f 20r) is also fairly unusual for modern wrestling, since the arm is pulled to the outside, rather than across the opponent (see HTK, f 63v). In addition, the elbow is levered, a detail more commonly found in arm drags in catch wrestling (see also Table 5). These aspects are, however, present in the inside 2-on-1 arm tie (see Table 3), which can also be used aggressively to pull the opponent, so it might be more appropriate to refer to it as such.

Judo terminology offers more precise and internationally understood distinctions when it comes to blocking and hooking the legs (Table 2). However, this is not always an advantage here, because some techniques fall between the prescribed execution, and a number don't fit well into any. For example, most of the outside hooks—referred to as 1st (f 17v), 5th (f 33v) and 6th (f 34r) "*twirch*" in BF[37]—could most appropriately be classified as variations of *Sukui Nage* in Judo, others as *Kosoto Gari* or *Kosoto Gake* (the distinction being difficult based on the descriptions and depictions), at least four variations could fall into a variety of categories. Notably, the classic hip throw (*O Goshi*) is not depicted, only variations where the opponent is thrown over the leg.

The only technique from this list that is lacking in modern wrestling but present in Judo is the *Tomoe Nage*, since in order to execute it, the attacker has to fall on

[34] Not very differentiated descriptions, so a variety of finishes—sweeping, hooking, blocking and sagging (aka "Polish throw")—are possible.

[35] MYSNYK ET AL. 1993: 36–37.

[36] See ANDREJEW & TSCHUMAKOW 2008: 81.

[37] Other treatises refer to these as "*fur treten*" (HTG, f 119v), "*hinder tretten*" (HTM, f 55v), and "*Schlossringen*" (Auerswald).

Technique	Folia in BF	Total
O Uchi-Gari	16r, [16v]	2
Sukui Nage	17v, [18r], 33v, [34r], 50r, 52v, 59r, 64v, 65r	9
Kosoto Gari/Kosoto Gake	16v, 34v, 54r, 70r	4
Unclear:[38] *Tani Otoshi/(Nidan) Kosoto Gari/Sumi Otoshi/Sukui Nage*	41v, 42v, 45r, (61v)	4
Osoto Gari/Osoto otoshi	18v, [19r], 20v, [33r], 69r	5
Osoto Guruma	19v, [20r]	2
Kuchiki Taoshi (illegal)	44v, 63r, 66v	3
Soto Makikomi	35r	1
Morote Gari (illegal)	42r	1
Tomoe Nage	41r	1
Ippon Seoi-Nage	35v[39], 38v, 39r, 39v, [40r], 60r, [60v]	7
Kata Guruma (illegal)	36r, [36v]	2
Koshi Guruma (variation: *Kubi Nage*) (illegal version)	49v, [50r], 50v	3
Tai Otoshi	43v	1
Unclear: *Kawazu Gake* (illegal)/*Uchi Mata/O Uchi-Gari*	(17r), (46v), [47r], [47v], 48r, [48v], [49r], [49v]	8

Table 2: Judo: throws and takedowns.[40]

his back, risking giving up points or even a pin in wrestling. It is however present in medieval wrestling and identical to Talhoffer's "*Buben Ringen*" (HTM, f 105r).

Some techniques such as *Kawazu Gake* were banned in Judo several decades ago and are therefore not routinely taught. Others, such as *Kuchiki Taoshi*, *Morote Gari*, the shown variations of *Kata Guruma* and *Koshi Guruma* as well as all versions of *Sukui Nage* with a leg grab were banned from competition in 2013 and are therefore not taught or used in every club, either.

It should be noted that BF doesn't actually depict any throws from grips on clothes, only grabs that are defended (see Table 4); therefore, all the takedowns shown would be the no-Gi version of the listed Judo techniques.

One aspect that distinguishes BF among other treatises is the description of tie-ups (grips) and advantageous positions from which to continue wrestling. In Table 3, I have included both the starting tie-ups as well as tie-ups which are used as the *Stuck* develops. Again, we can fit most of the tie-ups into wrestling terminology,

[38] These techniques are hard to categorize, since the description is somewhat vague.

[39] Possible finish of Seoi Nage; see KIMURA 1974: 34–35.

[40] Terminology based on DAIGO 2016 and INOKUMA 1979, unless otherwise noted.

Technique	Folia in BF	Total
No tie up (shooting distance)	42r, 42v, 70r	3
Single inside tie	69v	1
Elbow and inside tie	16r, 17v, 18v, 19v, 20v, 33v, 35v, 36v, 39r, 39v, 40v, 41v	12
Collar tie	70v	1
Reverse collar and elbow	34r, 34v, [35r], 71v	4
Over-under	43r, 43v, 44r, 44v, 45v, 46r, 46v, 47v, 48r	9
Inside 2-on-1	19v, 20r, 33r, 45v, 68r	5
Outside 2-on-1 (illegal variations)	36r, 40v, 65v, 68v, 69r	5
Outside 2-on-2	56r (?)	1
Single overhook	66r	1
Double overhook	53v	1
Overarm bearhug	[61v]	1
Overarm bearhug (rear)	[62r]	1
2-on-1 overhook	44r, 54v	2
Single underhook	43r, [44r], 44v, [45r], 46v, [47r], [48r], 48v, 49r, 50r[41], 50v, [66r]	12
Underarm bearhug	[41v], [51v], [53r], [53v], [54r], [61r]	6
Underarm bearhug (rear)	[57r], [58r], [59r], [59v]	4
2-on-1 underhook	43v	1
Lat grip over the back	[47r], 48r, [48v], 51r	4
Hair grab (illegal)	[70v], [71r]	2

Table 3: Amateur wrestling: Tie-ups.

with the exception of the grips on the clothes and the hair, which would be illegal in wrestling, therefore they don't have specific names.

Some of the tie-ups that are frequently depicted—like the bearhugs—are never explained, since only counters against them are shown; in fact, almost all the depictions show a very suboptimal double underhook placed in the middle of the back—where they are easiest to counter—rather than correctly either high on the shoulders or on the lower back.

The inside/biceps tie appears most frequently. The modern inside tie is usually thumbless—hooking the biceps with the wrist and gripping/pulling the triceps with the long fingers and the aligned thumb in a "monkey grip"—whereas the medieval grip is located on the biceps, thumb on the inside, long fingers on the outside. This grip is riskier in pure wrestling, since the obvious counter is to try to dislocate the thumb, but safer when weapons are involved, since it allows for a more secure

[41] Depicted wrongly as an overhook.

control against thrusts[42] as the medieval inside/biceps tie is actually primarily a push rather than a pull.

The inside 2-on-1 is featured fairly often; it is still quite popular in the modern Nuremberg school of Greco wrestling, but more rarely seen internationally than the outside 2-on-1. It should be pointed out that the attacker is always at risk of giving his back to the opponent in this tie, therefore it is recommended to place the inside elbow in the opponent's chest to prevent him from going around (see HTK, f 49v).

The tie-up on 56r is not easy to categorize from the description, since we aren't told precisely what to do with the left arm. In my opinion, it should be fed through the opponent's arms, overhooking his right arm before gripping the left forearm, making it an outside 2-on-2, a fairly rare and hard to counter variation found sometimes in the Caucasus wrestling traditions, usually resulting from attempts to counter an outside 2-on-1. If the left arm simply grips the opponent's left forearm without the overhook, the tie-up would be inherently risky, since the defender would pretty much give up his back, especially when standing sideways as depicted (he would have to walk backwards faster than his opponent can walk forward to prevent this); on the other hand, once a 2-on-2 is in place, both of the opponent's arms are effectively neutralized, meaning he can be thrown almost at will.

Another unusual variation is the reverse collar tie (ff 34r–35r; f 71v). This is uncommon in modern wrestling because it invites the elbow pass shown on folio 35r. It is, however, used in traditional Thai boxing and Burmese *Lethwei*, since it offers great protection from head butts; overall, it is a version that creates more distance, while the conventional collar tie (f 70v) is better for pulling the opponent in.

The grips on clothes can be categorized according to Judo terminology (Table 4). There is a clear preference for gripping the lapels, preferably high on the collar;

Table 4: Judo grips and locks.[43]

Technique	Folia in BF	Total
Lapel grip	[56v]	1
Double lapel grip	[56r]	1
Cross lapel grip	[55r], [55v]	2
Low lapel grip	[55r], [55v]	2
High collar grip	[62v], [63r], [63v]	3
Low Gi grip (hips)	[54v]	1
Waki Gatame	18r	1
Ude Gatame	38v, 39r, 43v (illegal variation with shoulder lock), 60r, [60v], 62v	6
Ude Garami	68r, 68v, 69r, 71r, 71v, 72r, 73r	7

42 I would like to thank RALF SCHEPPERS (Münster/Germany) for pointing this out to me.

43 Terminology based on INOKUMA & SATO 1979 and KIMURA 1974.

sleeves don't seem to be gripped, and belt grips are also lacking. Interestingly, all the techniques described in jacket wrestling are breaks and counters. It appears the author deemed it important to be able to defend against such grips, but didn't think they needed to be learned.

A number of the elbow locks can likewise be categorized in Judo terminology; where they can be, there is a slight preference for bent-arm locks (*Garami*) over straight arm locks (*Gatame*). Most of them are used in throws rather than on the ground, which is illegal in modern Judo for safety reasons—they give the defender basically the option to jump into the throw faster than the lock comes on, or to suffer at least a dislocated elbow. A number of elbow locks are however hard to place into a Judo framework, since they arise from situations that aren't meant to be held semi-statically, but rather simply ripped through.

Catch wrestling terminology is looser and allows us to categorize more techniques (Table 5). We can clearly see a preference for targeting the elbows (26 times total) over the shoulders (4), fingers (4), neck (4, including the flying mare), knee (2, including a counter) and wrist (1). Again, in the vast majority of cases, the locks, gouges and fishhooks are used to make a takedown, rather than as an end in them

Table 5: Catch wrestling.[44]

Technique	**Folia in BF**	**Total**
Flying mare	57r, [57v]	2
Neck crank	53r	1
Camel clutch (neck crank variation)	73v	1
Figure 4/double wrist lock	69r	1
Elbow crank (various bent elbow locks)	40v, 44r, 54v, 55r, 56r (?[45]), 56v, 63v, 66r, 68r, 68v, 71r, 71v, 72r, 73r	14
Straight arm lock	18r, 20r, 36r, 38v, 39r, 53v (?[46]), 60r, [60v], 62v, 65v	9
Shoulder lock	43v, 49r	2
Surfboard	74r	1
Chicken wing/turkey wing with shins	73r	1
Straight knee bar	58r, [58v]	2
Wrist lock	55v	4
Single finger lock (illegal)	59v, [60r], 72v, 74r	4
Gouging (illegal)	47v, 51v, [52r], 61r, 62r	5
Fishhooking (illegal)	45r	1

[44] Terminology based on BEAUMONT 1998.

[45] Not mentioned, depends on the execution of the technique; not included in the count.

[46] As above.

Technique	Folia in BF	Total
Right cross/right straight/hay-maker	[64r], [64v], [65r], [65v], [66r], [66v], [68v]	7
Right slap/hammer fist	[67v], [68r], [69r], [69v], 70r	5
Left jab	66v	1
Left uppercut	70v	1
Right shovel hook	66r	1
Left parry	65r, 68r	2
Right parry	64r, 64v, 65v, 68v, 69r, 69v	6
Slip	66v	1
Kick	60v (to back), 67v (to groin)	2
Knee to groin (illegal)	61v, 67r	2

Table 6: Mixed Martial Arts.

selves. The only exceptions are the come-along (f 72v) and the holds (ff 72r, 73r–74r), where pain and threat of joint breaking are used to control the opponent(s).

The punches, parries, evasions (slips), kicks and knees can be categorized with modern MMA terminology (Table 6), which is based on boxing. Interestingly, the opponent seems to attack with just two punches, a right haymaker and a right slap or hammer fist. The defender favors a more varied arsenal also featuring a jab, an uppercut and a shovel hook. These are always done in "dirty boxing" or "holding and hitting" fashion, using a tie-up or going for a takedown at the same time. The parries also lead to takedowns and/or joint locks.

It is worth noting that parrying right punches with the right hand is discouraged in most schools of modern boxing and MMA, since it opens you up to a counter-punch with the left. However, as explained on folio 64r, the anonymous author advocates a cross arm guard with the right arm on top—a good choice for an in-fighter looking to close the distance—from which it is not unusual to block a right-hand punch with a right-hand block. Still, it is interesting that the techniques in the dirty boxing section lack the usual "this is possible on both sides" and that not a single left-handed strike is shown.

While we will not evaluate the techniques from the older unarmed section of BF in the same detail, we will still have a short look at them (Table 7). Since all verbal instruction is lacking—like in the contemporary *die Blume des Kampfes* (FLW)—the only way for the plays to be used in training would be under the supervision of someone familiar with them—even more so since WELLE points out that the depictions are often not entirely correct in terms of biomechanics.[47] There are eight techniques in total, two variations of the fireman's carry, two inside singles, one outside single, one outside hook, one outside trip and a spear double. Folia 98v and

[47] WELLE 2014: 92–95.

99r (left) show attacks with one knee on the ground, something we don't see in the newer section of the treatise but frequently in modern grappling ("shot"). Interestingly, only two technique variations—the fireman's carry from 99r and the spear double from 100v—are also featured in the newer section, the others are present but in somewhat different variations.

Also, all the techniques carry over well to armored fighting, as can be seen on folio 81v, where the reverse lift/reverse inside single appears, on folio 93v, where a variation of the spear double using the sword is shown, and on folio 94r, where the outside trip (also shown on folio 90v and on folio 98v with the large shield) and second variation of the fireman's carry are depicted in the context of longsword fighting.

However, the armed section also includes an X-(b)lock on the forearm with the dagger (f 83v), an inside single with the sword grip (f 84r bottom) and one with the blade (f 83r top), an arm wrench (f 86r), a crotch lift gripping the wrist through the legs (f 88r bottom), taking the opponent's back via *Zucken* and proceeding into a takedown (ff 89r, 93r bottom), a *Sukui Nage*/reverse lift combined with an outside single (f 93r top), two variations of the outside single with a dagger (ff 88r top, 94v), two outside singles with the sword (ff 85r bottom, 90r bottom), at least one outside hip throw (f 102r bottom, probably also f 85r top), one outside trip with the shield (f 98r), six outside hooks using the sword as grappling aid (ff 81v top, 84r top, 90r top, 90v bottom, 102r top, 104r[48]), two "broomstick" snapdowns blocking the opponent's right leg with the left (f 87r as a counter to a lock/disarm with daggers, f 93v bottom using the sword blade), a snapdown (?) with a neck crank from an X-lock using a dagger (f 103v), an admirably vicious version of a grapevine throw (*Kawazu Gake*) while punching the opponent in the face with the shield rim (f 104v) and a pin with an inside "Turk" to facilitate stabbing the downed opponent

Table 7: Unarmed wrestling in the older section of BF and analogous techniques in wrestling and Judo.

Folio in BF	Technique(s)
98v	Bottom: Kneeling inside single/*Kuchiki Taoshi* from reverse collar tie
99r	Top: Fireman's carry/*Kata guruma* Bottom: Kneeling outside single with outside hook and forearm over chest/*Sukui Nage*
99v	Top: Reverse lift with inside leg grip/Reverse inside single (right arm inside) Bottom: Outside hook/*Kosoto Gari* from cross grip lapel grip/collar tie
100r	Top: Fireman's carry/*Kata Guruma* with outside leg grab Bottom: Outside trip/*Osoto Otoshi* from over-under
100v	Spear double/*Morote Gari*

[48] A particularly clever version using the sword blade as a lever between the legs.

in the face with a dagger (f 103r). Whoever the author of the older section was, unlike the author of the newer section, he apparently wasn't taking any prisoners.

Not only are the wrestling techniques used with great diversity in armor, but the repertoire is actually expanded significantly in the armed section with additional techniques—to at least 30 variations of takedowns and one follow-up—unlike the newer section, where the unarmed plays are in the majority. In case the older section, the wrestling techniques might simply have been intended as training exercises for armored fighting, unlike the newer section, where they are used in various unarmed scenarios. It might also indicate that these were considered the basic wrestling techniques used in this particular school of fighting.

3.2 GAPS IN *BAUMAN'S FIGHT BOOK*

One obvious gap in BF is that (unlike in most other treatises covering wrestling) the basic wrestling stance is not depicted as a starting position; we can only guess whether the author favors a square stance (feet parallel, 1–2 shoulder widths apart) or a stagger stance (one foot closer to the opponent); we can only observe the wrestlers already in motion. Generally, the stances depicted are fairly upright for modern Freestyle standards, but somewhat bent over for Judo standards. From a modern wrestler's perspective, it looks like a Greco wrestler attempting to wrestle Freestyle. This approach makes sense under the premise of being useable in armed fighting and with punching involved—the stance in modern MMA is similar.

In terms of movement, one clear difference between BF and modern wrestling is the way of stepping when attacking: while modern wrestlers would usually hit a knee on the mat (a.k.a. "shot") when moving under the opponent in leg attacks, fireman's carries, and shoulder throws in order to attack lower, we don't see that in BF, probably because the concept only works well on soft ground or in armor.[49] *Ludwig von Eyb* (LE), on the other hand, suggests hitting a knee at least occasionally, and so does the older section of BF (see Table 7). Instead, the wrestlers in the newer section of BF always apparently attack with a fairly big step, visually exaggerated by the pointed shoes worn. It is clear that they lead in these steps both with the right and the left leg, which indicates that they either start from a square stance or routinely switch lead legs in a stagger stance.

These wide steps and stances—especially when doing an outside hook—make a wrestler fairly vulnerable to grapevines and inside hooks (*Haken*). Monte does not even mention the outside hook,[50] but his translator discards it as "inept".[51] If we look at treatises from the end of the 15th and early 16th century—when variations of the Haken seem to have become more commonplace—in fact wrestlers seem to have adopted a narrower stance. While the Haken can be countered in a number

[49] Judo, which uses harder mats than wrestling, has adopted this concept very hesitantly, and old-school lines still disapprove of it.

[50] "*trascorbada*"

[51] FORGENG 2018: 372.

of ways, that is tricky against a *Haken* specialist. The author of BF does not give any detailed instructions on the *Haken* in the surviving pages and on folio 46v even writes that it isn't worth it[52] to mention the possible variations. He therefore appears to assume that the reader is already familiar with the technical options that arise from the basic *Haken*—possibly from oral instruction, as the quote "as you have been instructed before"[53] may suggest—and therefore apparently sees no need to elaborate. Other authors such as LE, on the other hand, show a large number of possible—often fairly complex—variations.

Compared to other contemporary treatises, we can also observe that the variations of the duck-under to lift (*Te-Guruma* in Judo)—a common counter to fireman's carries, etc.[54]—are notably absent in BF.

3.3 TEXT-PICTURE DIFFERENCES AND TRICKY INTERPRETATIONS

There seem to be a number of deviations between the text and the pictures, as RAINER WELLE has pointed out.[55] To point out the more consequential ones: on folio 47v, one wrestler both does the *haken* and the prescribed counter to it; on folio 49r, the thigh of the right wrestler is missing, the calf is colored incorrectly; on folio 50r, the wrestler on the left is shown with a left overhook rather than an underhook; on folio 56v, an unusually early stage of the technique is shown and therefore the picture is unclear; on folio 58v the wrestler on the right seems to pull the opponent's leg to his side, rather than between his legs, and on folio 73r, the text describes holding one prisoner, chicken winging his left arm with the left shin, while the picture shows two prisoners, turkey winging the right arm with the left shin and vice-versa. Other, smaller mistakes include anatomically incorrect depictions of the hands.[56]

The most obvious mistake is that the description from folio 18v describes the depiction of folio 19v and vice-versa, as noted next to the pictures; the picture on folio 19v also doesn't show gripping the opponent's left elbow, which is essential to the technique, more so than the armpit grip that is shown. In fact, from the position as it is depicted, the attacker in red likely would be the one to be thrown, since he is too upright—all the other wrestlers would have to do is to lean forward. WELLE dismisses the re-counter of this technique on folio 19r as depicted entirely wrong, stating that the correct version is shown on folio 43r.

---

[52] "*pillich*"

[53] BF, f 46v: "*alstu dan vor vnter weist pist*"

[54] See e.g. HTK, ff 51v, 52v, 59r; GF, ff 38r, 47r, 52r.

[55] WELLE 1993: 114–122; WELLE 2014: 225–229.

[56] See WELLE 2014: 225–229. In addition, I would argue that on f 69v, it would be highly advisable to either also put in an outside hook with the left leg, or step on the opponent's right foot with the left (neither described nor depicted), otherwise he probably would not fall reliably if he's not completely out of balance from the punch.

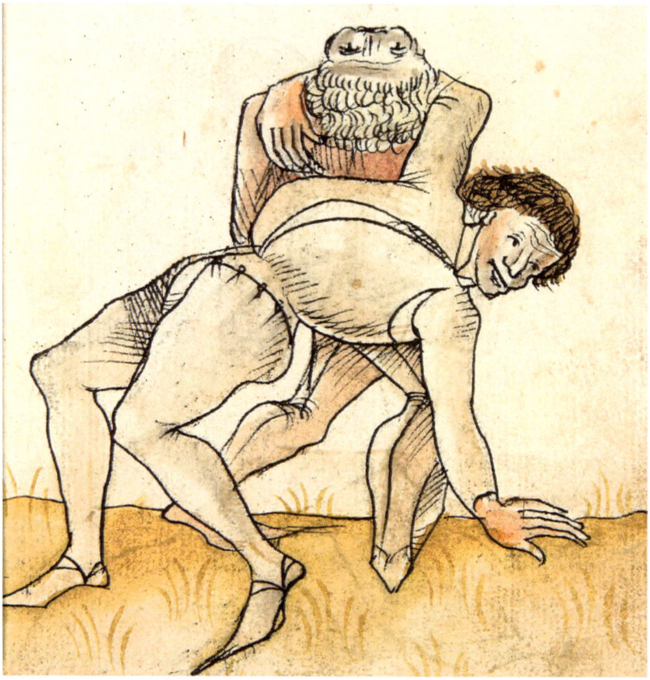

Figure 1: Rare techniques in BF, the falling arm throws on 19r (left) and 43r (right).

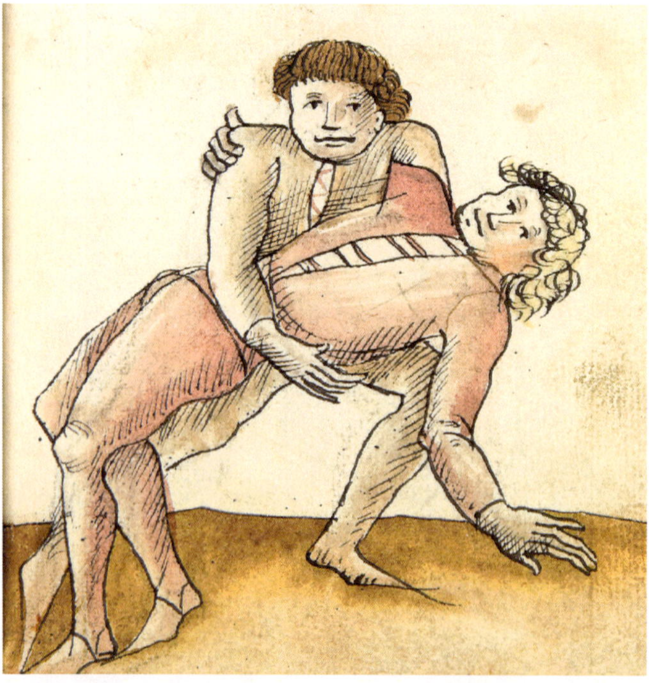

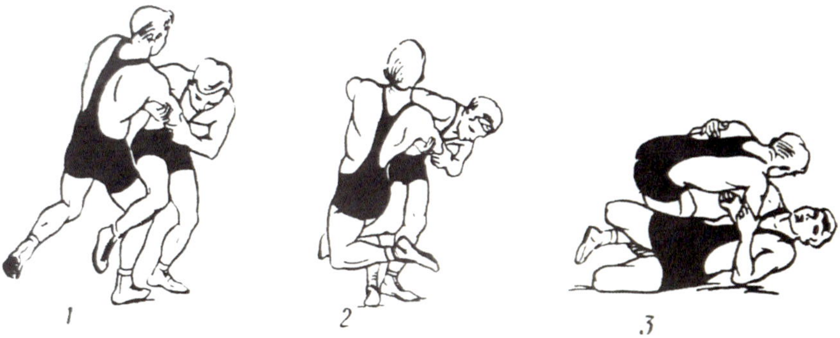

Figure 2: A technique similar to the one shown on BF f 19r, here as a counter rather to an outside trip rather than a re-counter. The wrestler doing the counter is on the left, the end—a pin with an outside Turk—is not shown. Note the jumping back step of the left wrestler from picture 1 to 2.[57]

The related technique on folio 43r is also rather unusual—both today and in medieval treatises[58]—and terminologically not easy to reference. FOELDEAK shows a similar technique and calls it "*Verkehrter Armfallschwung*" ("reverse falling arm throw"),[59] and COSNECK shows a picture sequence of it as a counter to a belt grab (Fig. 3), but doesn't name the technique. I personally haven't seen the technique in use or being taught; however, based on the descriptions and depictions, the technique involves a 270° turn and ends like a kneeling hip throw; however, this is achieved by bending the torso back at the waist while turning in, making the technique overall more similar to an arm spin.

Figure 3: An almost identical technique to the one on BF f 43r, here as a counter to a belt grab.[60] A backstep is shown in frames 2–3, but stepping the right foot over and putting the right knee down or stepping through the opponent's legs—as in an arm spin—would be possible if the hips of the bottom wrestler in frames 1–2 were higher (as on f 43r).

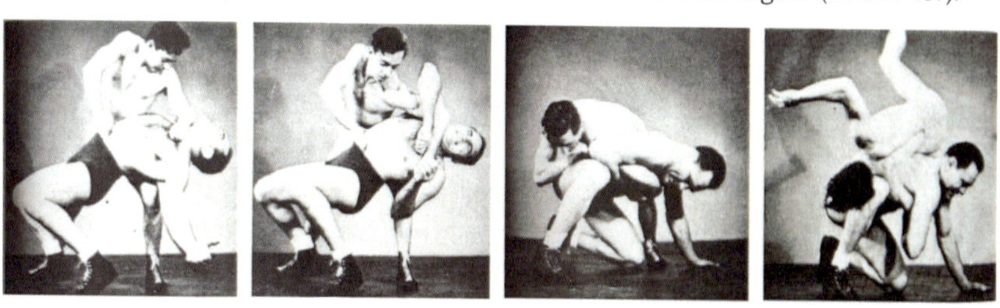

[57] CHINAZZO 1970: 96, fig. 84.

[58] A variation of it called "*der dritte fueß*" appears in GF on f 37v.

[59] FOELDEAK 1952: 60–61.

[60] COSNECK 1944, 47

I initially listed folio 71v (fig. 4) as an outside hip throw (as seen in HTK, f 50v). However, when coming across Paulus Kal's depiction, I concluded it must be very similar to a technique popular among Russian wrestlers today, the Saitiev/Sidakow slide-by. BUVAISAR SAITIEV used it to take the opponent's back, while ZAURBEK SIDAKOV keeps his front foot between the opponent's feet and proceeds to an outside hip throw with an inside sweep (see fig. 4 on the bottom). SIDAKOV's version is actually pretty close to the one on BF, except that he uses a regular collar tie, while folio 71v shows a reverse collar tie. Also note that today's wrestlers tend to lean to the side while pulling the elbow high to allow the opponent's arm to slide off; otherwise the forearm can get stuck and an inside elbow wrench will be the result, which is likely the case here.

Another tricky case is folio 55r (fig. 5, which is labeled a "concealed technique",[61] and "written so that not everybody understands it"). As WELLE points out, left and right (and vice-versa) were switched in the initial version and later corrected.[62] However, with or without corrections, the picture and the text don't match

Figure 4: Slide-bys through time: BF, f 71v (right, top); PKM, f 92r (right, bottom), and ZAURBEK SIDAKOV at the 2019 World Championship (below).[63]

[61] "*verporgens stuck*"

[62] WELLE 2014: 227.

[63] https://www.youtube.com/watch?v=zRXZYuFcA3c, 0:00 – 00:10 sec (27.05.2022).

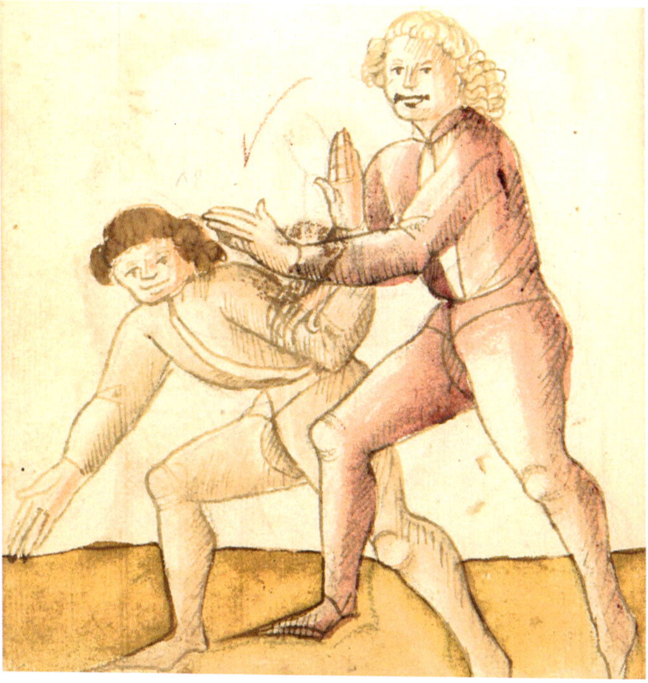

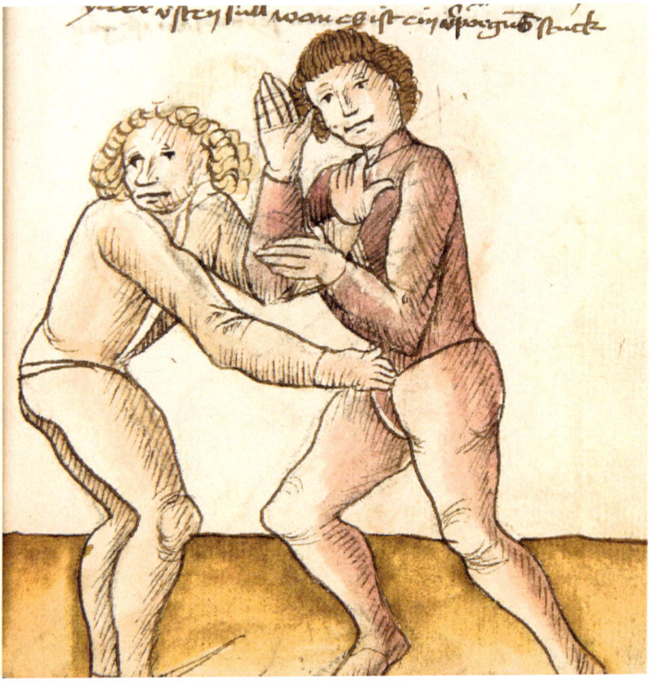

Figure 5: BF, f 55r (top), GF, f 56r (bottom).

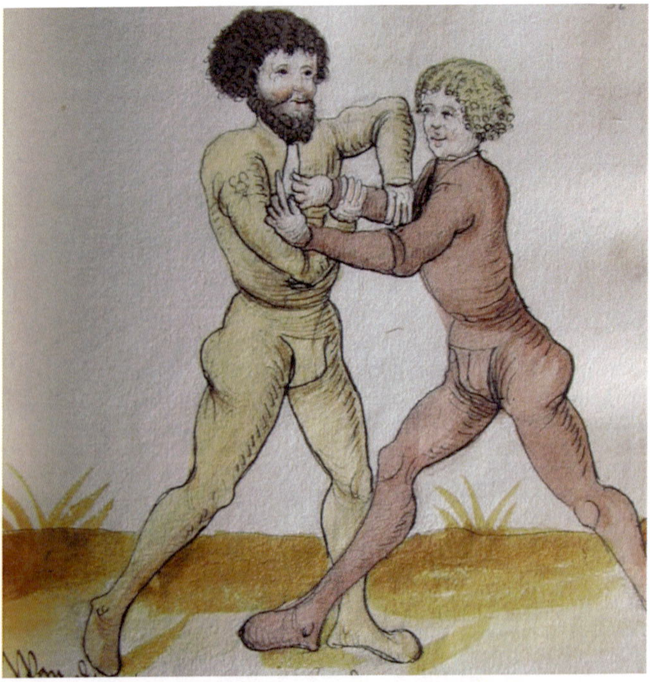

completely. In fact, following the uncorrected description, one could end up with one version of an elbow lock (f 68r), following the corrected version another (f 68v), while from the pictured position, at least one more variation would be possible. Deciding which one was meant here is not easy, but the good thing is that they all work. A possibly corresponding technique is found in GF (f 56r), which seems to be written more clearly and indicates that the technique is most likely another inward elbow crank.

3.4 DO I NEED ALL OF THEM?

The number of techniques a style or school teaches doesn't necessarily reflect the number of techniques a practitioner will perform in a match or fight, nor does the number of variations taught for a certain technique indicate an individual preference for that.

While grapplers are usually encouraged to learn all the techniques of their school and evaluate them for their own use ("passive repertoire"), they will usually limit themselves to a handful of techniques ("active repertoire" or "favorite techniques" in wrestling, *Tokui-Waza* in Judo) in competition. These should fit the grappler's style and physique and ideally combine well.

These techniques usually make up the bulk of training: YURIY SHAHMURADOV recommends spending 50% of training time on developing the "most favorite holds".[64] He elaborates that a wrestler should work on 2–3 favorite techniques per session, training each of these techniques 3–4 times per week (thereby implying that a wrestler should have 5–8 favorite techniques).[65] He stresses the need for constant improvement of these techniques in their application in various scenarios and lists a number of strategies employed to achieve this as well as overcoming loss of trust in favorite holds after repeated failure to execute them under stress. These include (but are not limited to) performing training exercises specific to the technique, varying training partners, varying the speed of execution, training pre-fatigued and performing the technique against a training partner expecting it and tasked with defending against it.[66]

[64] SHAHMURADOV 1996: 16.

[65] According to my coach RAMIS, the active repertoire of a top-level wrestler should include at least 2–3 techniques on the feet and 2–3 techniques on the ground. One should aim for 800 full repetitions—including set-up, entry, execution, follow-up—of a technique per training session. In Judo—where the grips on the jacket are more repeatable than grips on a sweaty person and therefore require fewer constant adjustment—this is often done in the form of resisted entries (*Uchi-Komi*) with a partner or rubber band rather than full throws, saving the partner from the repeated impacts and allowing for up to 10.000 repetitions per day. See ISHIKAWA & DRAEGER 1961: 177–189 and STARBROOK 1978: 40–42, 62–63.

[66] SHAHMURADOV 1996: 12–16.

4 Didactic Structure

Methodically, the newer *Ringen* section of BF uses a combination of a single picture for each technique, combined with 5–10 lines of descriptive text. The volume of text describing the plays is higher than average for a mid-15th century treatise. Techniques are broken down into phases via text[67] rather than through a sequence of pictures, unlike modern grappling books where the pictures usually dominate over the text, and every instructional step usually gets its own picture. Usually, after describing the initial situation, 3–6 phases are described (compared to typically 4–8 in modern instructional books). In some cases, important points are highlighted, implying common mistakes,[68] sometimes additional tactical advice is also given (e.g. f 18r).

While it can be argued that someone familiar with the subject can potentially draw all the necessary information from text, including information that is close to impossible to transmit in pictures or video clips without audio, good depictions still serve a number of valuable purposes. They can help to find a technique faster, clarify terminology, show unmentioned details (like exact position of the feet, etc.) and be powerful visualization aids that ideally make an experienced viewer "feel" the technique, almost as if watching the entire motion, leading to a more vivid, less mentally tiring reading experience.

In terms of clarity of the depictions, the newer *Ringen* section of BF addresses several problems that single-picture depictions of a technique bring with them. First, the wrestler executing the technique is usually[69] dressed in red.[70] The painter of the wrestling section also captured the dynamic of the weight and energy transfer in the techniques well, arguably surpassing Dürer's copies in AD,[71] which seem to somewhat lack the dynamic of the original—maybe because he copied depictions, rather than painting moving wrestlers. In order to transmit this dynamic, the standard system in the 20th and 21st century is to show a sequence of 4–8 photos or drawings per technique (see Fig. 4 and 5), sometimes including arrows to indicate direction of force[72] or even a QR code to a complementary video file.[73]

While the order in which the wrestling techniques are presented may seem arbitrary at first glance, Rainer Welle suggested a functional grouping of the techniques of BF in eleven groups (Table 8).

---

[67] See Welle 2014: 117–126.

[68] For example, ff 17v, 18v, 19v, 20v, 34r, 36v, etc.

[69] Except on ff 18r, 19r, 37r, 40r, 41v, 47r, 50r, 52r.

[70] Fiore achieved this by depicting figures with garters and/or crowns, so the idea was not entirely new.

[71] For the comparison see Welle 2021: 125–185.

[72] E.g. Rothert & Tepper 1988; Rothert 1994; Andrejew & Tschumakow 2008.

[73] Ruch et al. 2014.

Group	Folia	Situation[74]	Situation (translation)
A	16r – 41r, 42r – 42v	*Zulauf aus langen Armen*	making contact to inside and outside tie-up with straight arms
B	41v, 43r – 50v	*Ringen aus dem Gleichfassen = Zwiegriff*	wrestling from over-under tie-up
C	51 r – 54r	*Ringen aus dem Untergriff von vorne ohne Armein-schluss*	wrestling from underarm bearhug
D	54v – 56v	*Ringen mit Kleiderfassung*	wrestling with grips on the clothes
E	57r – 60v	*Beidseitiges Fassen von hin-ten ohne Armeinschluss*	underarm bear hug from behind
F	61r	*Beidseitiges Fassen von vorne mit einseitigem Armeinschluss*	over-under tie-up[75]
G	61v	*Beidseitiges Fassen von vorne mit beidseitigem Armeinschluss*	overarm bearhug from the front
H	62r	*Beidseitiges Fassen von hin-ten mit beidarmigem Armeinschluss*	overarm bearhug from be-hind
I	62v – 63v	*Ringen mit Kleiderfassen*	wrestling with grips on the clothes
J	64r – 70r	*Faustschläge*	punches
K	70v – 74v	*Raufen*	brawling

Table 8: Didactic Structure of BF according to RAINER WELLE.

After thoroughly examining the layers in the book's binding, WELLE concludes that this order is well-structured, logical (except for bifolia 41/47 and 42/48, which should be switched), suggests that it is intentional from the beginning and that the wrestling part is most likely complete in the sense of not missing pages within the surviving section.[76] In his earlier publication, he had argued out that some techniques like attacks from the *haken* were most likely included originally and suggested that Dürer might have had access to these missing pages featuring up to 38 *Stücke*.[77] He later points out similarities of the additional *Stücke* in Dürer with

[74] Transcription by WELLE.

[75] "*vmb vangen*" may suggest an underarm bear hug from the front (compare f 57v).

[76] WELLE 2014: 31–56.

[77] WELLE 1993: 122–123.

GF.[78] However, in a number of these plays,[79] the text in Dürer's work seems to mirror the phrasing typical for BF, indicating that he either had access to additional pages from the same anonymous author,[80] or re-phrased the instruction from GF accordingly.

The didactic analysis presented by WELLE, combined with his detailed analysis of the pages and binding is a model approach for the interpretation of wrestling and fencing treatises. Certain additional sub-divisions could be made within these groups without disturbing the overall order. Personally, I would sum up sections **E–H** as "countering advantageous tie-ups", refer to section **J** as "brawling" and include folia 70v–71v, while folia 72r–74v might more aptly be summed up as "moving and holding prisoners"—or, in the case of folio 74v, extorting valuables from them.

Departing from certain scenarios and showing appropriate techniques is a classic way of organizing a grappling book.[81] A second way would be to present the techniques in the order suggested for learning.[82] Judo also distinguishes throws into functional groups which should indicate and highlight the mechanics of certain throws.[83] In wrestling, a basic distinction is often made between techniques where both feet of the opponent are lifted off the mat ("throws") and techniques where that isn't the case ("takedowns"[84]). They can be divided into further in leg attacks and upper body attacks and biomechanically into "structural groups", [85] which are another way of organizing a book.[86] Modern grappling disciplines usually employ several of these strategies in combination when presenting techniques in book form.

---

[78] WELLE 2021: 125–185.

[79] Ff 78–81, 86–88, 90–93, 99–103, 106–109. See WELLE 2021: 164–180.

[80] WELLE deems it possible that there were additional pages, which were not included in the "final redaction" of the treatise (WELLE 2014: 52–56).

[81] E.g. HTM, esp. ff 85r–87r. A similar approach was more recently chosen by MYSNYK et al. 1994. Pietro Monte also includes sections where he elaborates which techniques are useful in which situation; however, he describes the techniques first and then refers to them in various contexts (see FORGENG 2018, esp. 57–59, 63–66).

[82] E.g. the five groups of the Gokyo in Judo. RUCH ET AL. 2014 structures techniques according to the age a wrestler should learn them. CHAMBINAUD 2004 assigns a color grading of difficulty to each wrestling technique, similar to the color belts of Judo, while presenting the techniques in structural groups.

[83] See DAIGO 2016.

[84] However, "takedowns" is also often used to sum up both categories, since many techniques categorized as throws can be executed as takedowns, and vice-versa. For this reason, I chose to not make a distinction in Table 1 above.

[85] RUCH ET AL. 2014.

[86] GAIN ET AL. 1980.

5 SUMMARY

With its combination of concise tactical advice, fairly extensive technique descriptions and dynamic illustrations as well as comprehensive didactic structure, the wrestling section of the newer part of BF is to be considered the most important surviving treatise on *Ringen* from 15th century Germany and may contribute to the understanding of other, less extensive treatises on the subject. Among the illustrated German manuscripts, only LE (1500) and GF (late 15th century, compiled in 1508) feature a few more-extensive technique descriptions, but lack the clear didactic structure of BF. In the rest of Europe, the only 15th century author going into more depth on wrestling is Pietro Monte, who elaborates on a number of aspects that BF doesn't mention, such as stance, footwork, training and rulesets.

Whether or not it is complete, the newer section of BF features an impressive amount of unarmed *Ringen* techniques and counters (including some fairly advanced ones), arguably the highest of all the treatises not drawing from multiple sources. It also stands out in a number of other ways, like the description of several tie-ups and the inclusion of a fairly large number of "brawling" techniques in addition to the "dirty wrestling" (joint locks, gouging, fish hooking) in part also featured in other treatises. The focus is on counter-offence since the counters outnumber the attacks 3:1. The most common outcome is a takedown that gives the attacker the choice to remain standing. The joint locks usually end in takedowns and mainly target the elbows and shoulders, while the opponent's nose is a preferred target for soft-tissue manipulation. In short, we can conclude that the anonymous author of this section was not only a skilled wrestler, but also a vicious brawler—a winning combination.

The author of the older section certainly also knew his combat wrestling, especially if we look at both the armored and unarmored section. His system may have been wrongly overlooked so far, overshadowed by the newer section, despite being surprisingly diverse and brutally effective, and including a number of technique variations not featured in the newer section.

We can also conclude that BF was aimed at readers who already knew how to stand, move, fall, and train, and most likely were already familiar with at least the basic attacks in *Ringen*, meaning they were at least intermediate grapplers, had direct access to an instructor[87] or were instructors themselves. Even today, grappling books for beginners are usually intended to refresh techniques already learned on the mat; learning new techniques from books or videos is recommended only for advanced practitioners.

If we want to reconstruct the *Ringen* of BF, we of course need to study and compare the treatises from the 15th and early 16th century, with a special emphasis on potential influences and later copies/interpretations (in this case GF and AD). However, in order to do so, practical knowledge of grappling arts still practiced today can be highly helpful in understanding the mechanics—as shown in section

[87] WELLE 2014: 126.

3, most techniques have modern analogues that can and should be studied to better understand the short notes in treatises.

Unfortunately, wrestling is usually treated as an optional "add-on" to weapons practice in HEMA today, and most clubs simply don't have members proficient enough to be training partners for advanced *Ringen*, let alone to coach it. It therefore follows that we need to work on this problem, possibly with the help of modern grapplers. Improving the level of grappling in HEMA should allow for better—and safer—pressure-testing of theories and interpretations, both armed and unarmed.

I'll see you on the mat.

Author's Thanks

Much of this is article is also based on my own experiences training and competing in wrestling, Sambo, Judo, Sumo, Ranggeln and Schwingen, and conversations with my coaches—especially Jochen Engelhardt (Nuremberg/Germany, Greco Roman style) and Ramis Abdulkadirov (Derbent/Russia, Freestyle)—and training partners. Thank you for everything, I will forever be in your debt.

Thomas Locker (1st Dan) and his Judo coach Frank Thiele (9th Dan) were kind enough to discuss the Judo terminology for this article with me. It has been an honor and a pleasure.

Finally, I would like to thank my old friend Harald Rotter for proof-reading this article despite the narrow deadline.

Jtm ob dir ainer faustn wild schlagen so vass
im dy hant zu samen an seine hals und stich
im dar durch als hie gemalt stet so maint
er man hab im den hals ab geschnitten und slecht

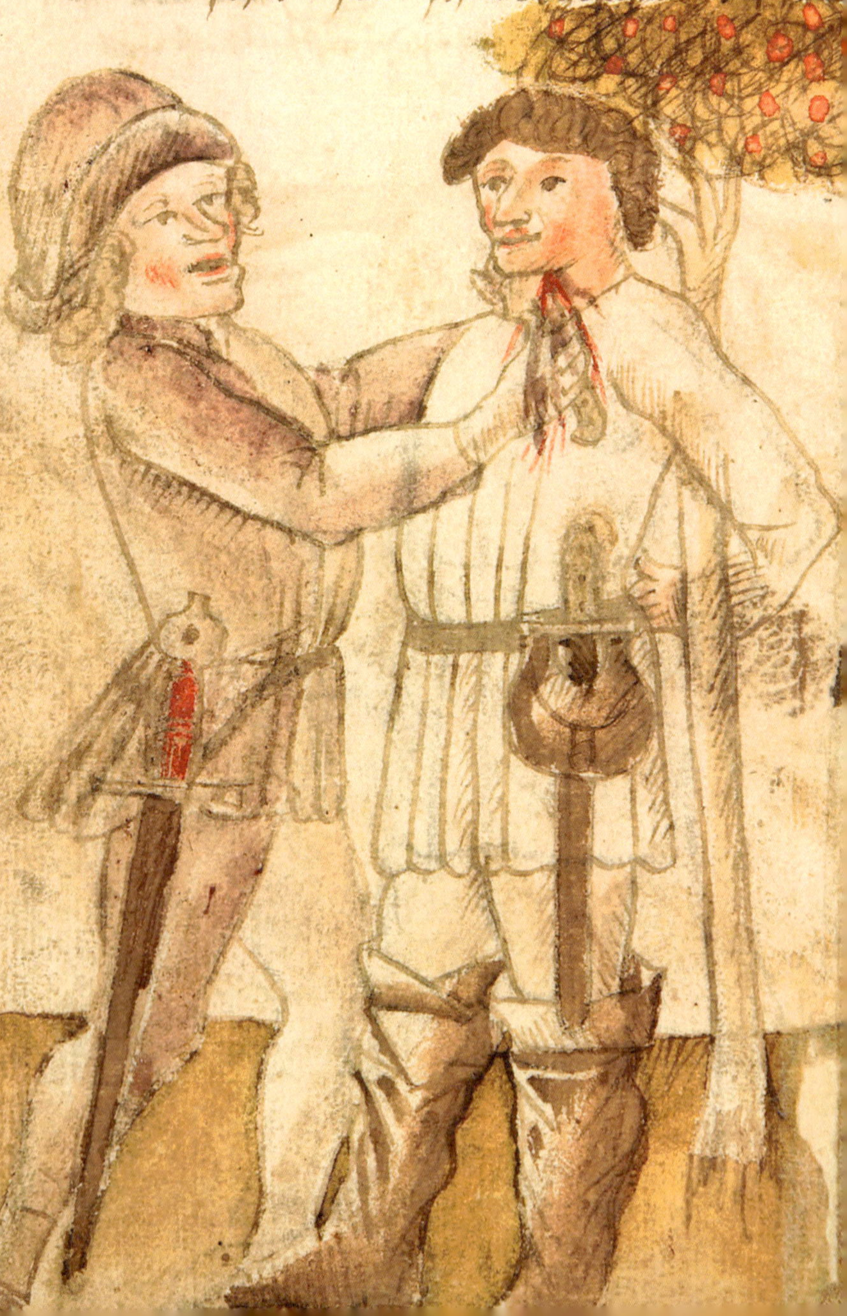

Nonlethal Techniques in the *Bauman Fight Book*: Context and Application

Charles Lin
Capitol Kunst des Fechtens (CKdF)

Marlene Hurst
Capitol Kunst des Fechtens (CKdF)

Bauman's fight book (BF) contains one of the most well-known medieval German treatises outside of the Liechtenauer corpus. The manuscript is a composite of three sections, the first part, which contains the sword techniques, also contains Messer and dagger techniques, with no indication the techniques are distinct bodies. Manuscripts containing fragments similar to BF (such as AD and AR) also do not give any special indication that the sword techniques are distinct from the Messer and dagger techniques. For many fencers that primarily study early Liechtenauer's long sword corpus, not only could BF be their first foray into Messer and dagger techniques, but it could also be their first exposure to techniques that are primarily nonlethal.

This article aims to provide a fencer primarily accustomed to Liechtenauer's long sword with an introduction to the "nonlethal" techniques in BF. For this article, we define "nonlethal" narrowly, to refer to techniques that contain locks, breaks, disarms, throws, and pommels. This avoids the more difficult problem of inferring the ultimate intent of a technique or the context in which it is used, but also forms a group of techniques which have unique training considerations for fencers.

We survey the techniques present in the manuscript, discuss training priorities, and contrast them with other fencing systems. We then provide some potential historical contexts in which BF might have applied. Lastly, we provide advice for safely and effectively practicing the techniques discussed.

1 Survey of Techniques

There are some complications in categorizing the techniques of BF. First, although the focus of this article is on "nonlethal" techniques, it is sometimes difficult to definitively categorize the intent(s) of a technique. For example, a hew to an opponent, can, depending on the target and the force involved, be intended as lethal or merely to wound. Similarly, many hews can use the flat of the blade for nonlethal application. For simplicity, we characterize techniques based on the mechanics by which the technique is concluded, rather than trying to infer intent—a hew, stab, slice, lock, break, disarm, or pommel, or the threat of an edge on the neck or head.

Further complicating this analysis is the fact that some techniques have compound actions with varying levels of lethality. For example, the sword technique

Figure 1: A sword-assisted throw (BF, f 10r).

in 10r uses the sword pressed against the opponent's throat to complete a throw (Fig. 1). While the throw appears to be the primary intent of the technique, it is entirely possible that a slice to the neck is implied. Inferring the intent of the technique might be application specific, and defies easy categorization into a lethal or nonlethal dichotomy. To simplify the analysis, we categorize 10r as both "Edge Threat" and "Throw." For techniques where the edge is laid against an opponent's face, head, or body and explicitly stated to push, draw, or move forward, these are categorized as a slice, and a wounder.

The vast majority of the techniques are nonlethal, with 65% percent of techniques being nonlethal or having a nonlethal component. Dagger and Messer techniques are overwhelmingly nonlethal (78% and 75%, respectively), and even the sword techniques have a focus on nonlethal endings that is unusual compared to Liechtenauer examples, with 54% percent being nonlethal.

2 Contexts for Nonlethal Techniques

Why are so many of the techniques from BF nonlethal? Part A of the text does not provide a context for the use of its techniques. Although Part C contains images associated with duels in harness and with specialized shields, it is difficult to associate these contexts with Part A because it is unclear when the separate treatises were originally compiled. The codicology of BF is outside of the scope of this article,[1] but we can provide several potential contexts where the nonlethal techniques might be useful.

First, many of the nonlethal techniques of BF remain effective even when the opponent is wearing armor, and, in some cases, are more effective. For example, shoulder hyperextension locks and breaks such as 22v and disarms such as 23r are effective against nearly all armors. Some types of armor can mitigate or prevent joint damage from elbow or wrist hyperextension locks such as 22v and 23v, but these techniques can still effectively neutralize an opponent, even if the joint itself

Table 1: Categorizing the techniques of Part A.

	Sword	Dagger	Messer
Total	26	14	8
Nonlethal	14	11	6
Lock/Break	2	5	1
Throw	6	4	2
Disarm	5	3	1
Pommel	2	0	4
Edge Threat	4	0	2
Wounders	10	6	3
None	2	0	0

Note: Because the categories are not mutually exclusive, the sum of the categories will exceed the total.
Nonlethal: Any technique that contains a lock, break, throw, disarm, or pommel to the head.
Edge Threat: Any technique that ends with the weapon pressed against the opponent's neck, head, or body with no indication to slice, stab, or otherwise break the skin.
Wounders: Any technique that contains a hew, stab, or slice, or suggestion to "work an opening."
None: The technique described has no concluding action, and is general advice for the fencer.

[1] See MICHAEL CHIDESTER's article in this volume (pp 1–11).

Figure 2: A joint-lock technique (BF, f 22v).

is protected. In addition, throws against armored opponents are usually easier than against unarmored opponents, given the higher center of gravity and reduced proprioception and balance.

Although the techniques of Part A show fencers without armor, the application of these techniques is possible against opponents in armor as well. For example, the dagger plays in folia 22v and 25v, while depicted without armor, also appear to show up in *Gladiatoria* with armor.[2] In this case, the label of "nonlethal" techniques can be misleading, because they can be used even if the ultimate objective of the conflict is lethal. For example, in Martin Huntsfeld's armored fencing

[2] BF, f 22v, is comparable to the technique in GW, f 43v. BF, f 25v, is comparable to the technique in GK, f 40v, and GK, f 42r.

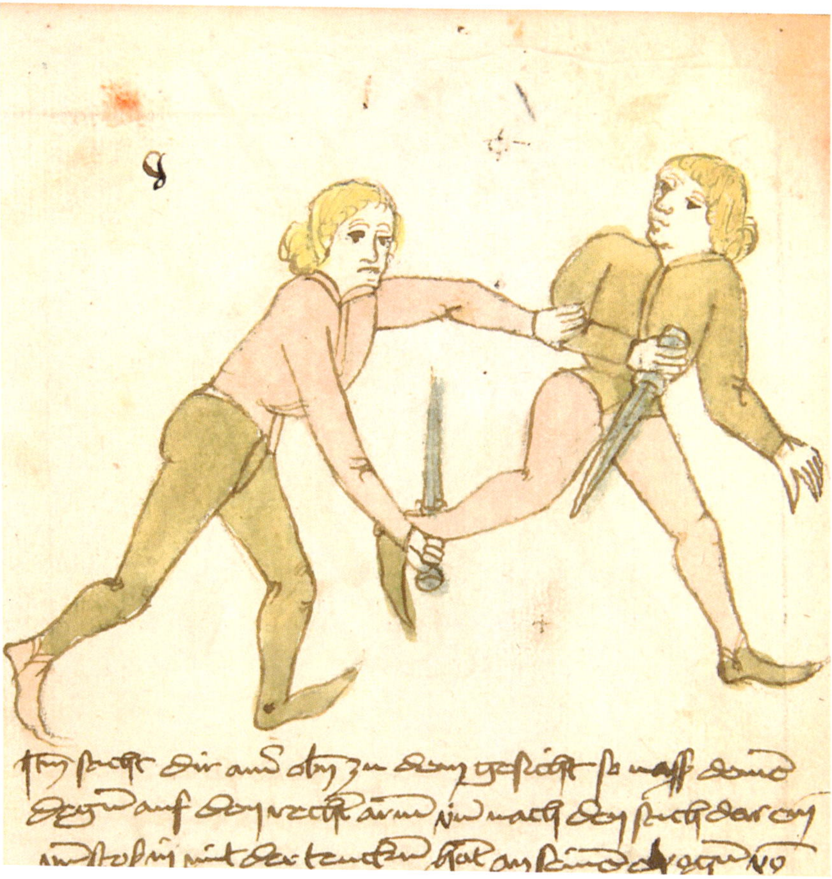

Figure 3: A dagger-assisted throw (BF, f 25v).

treatise,[3] the armored ground fighting plays assume the opponent has already been thrown to the ground, and the play begins with securing and pinning the opponent before using a dagger on them.

Second, BF was likely written at a time when noble violence was moderated by social and legal rules and norms surrounding feuds. In HILLAY ZMORA's examination of the feud in early modern Germany, he describes an "inimical intimacy" between feuding nobles, where conflicts are embedded in dense social, political, and financial networks. These feuds were shaped by the expectations of a moral community and generally sought to contain the destruction caused by conflicts.[4]

[3] See PD, L, PKW, HS, GF, AD, HC, G, PMD, PMM, PMW, and JMR.
[4] See ZMORA 2015.

For example, the Golden Bull of Karl IV in 1356 proclaimed,

> It is not allowed to attack someone, under the pretext of feud, with arson, robbery, or plunder unless the feud has been announced publicly to the one against whom one feuds, in person or at the place where he habitually resides, three days in advance, and this announcement be attested to by reliable witnesses.[5]

This passage suggests not only that the scope of activities within a feud might have been arson, robbery, and plunder, but that there were rules and norms that limited the damage of these property crimes. Indeed, in Georg von Puchheim's feud with the emperor Friedrich III over financial debts he was owed, Georg wrote a cartel of defiance declaring his feud and waited several months, writing to other German princes and hiring mercenaries, before deciding to escalate by kidnapping several people around Wiener Neustadt.

BF and its related manuscripts are associated with both Nuremberg and Augsburg, free Imperial cities that relied on a citizen militia, along with the aid of military professionals and allied nobles, for their security. These militias defended the city against external threats, such as the feuding nobles that surrounded the cities.[6] One famous *Raubritter* ("robber knight"), Gotz von Berlichingen, engaged in feuds against the cities of Cologne, Ulm, Augsburg, and Nuremberg, kidnapping nobles and merchants for ransom, which got him placed under the Imperial ban twice. Nuremberg was famous for sending patrols to capture enemies of the city to bring them in for trial under its own jurisdiction.[7]

According to ZMORA, feuds were "regulated by accepted rules of conduct and by a more or less fixed repertoire of sanctioned methods: sporadic yet organised, usually small-scale raids involving burning, looting, and causing all sorts of material damage to the rival and his interests. Brutality a *l'outrance* was on the whole exceptional. Killings were rare, and usually not premeditated."[8] In this context, the nonlethal techniques of BF could be valuable for these limited acts of violence, without risking the damage that an unsanctioned killing might cause to one's reputation.

The city militias of cities such as Augsburg and Nuremberg, with the aid of professional guards, also protected against internal threats—in addition to military and firefighting duties, these could include police or peacekeeping duties, such as patrolling the streets, detaining suspicious persons, and intervening in brawls and other altercations.[9] These duties extended not only to the persons on watch, but to all men of the city. According to Tlusty, "in communities with populations ranging from five to forty thousand, many of those being challenged or arrested by both

[5] ZMORA 2015: 34.

[6] See POPE 2017 and CHANDLER 2013.

[7] See POPE 2016.

[8] ZMORA 2015: 44.

[9] See CHANDLER 2013 and TLUSTY 2011.

Figure 4: Holding three men prisoner (BF, f 74r).

civilian and professional guards were likely to be friends, acquaintances, or even sworn enemies..."

Because the use of force within cities also occurred in dense social networks, as in the case of noble feuds, these uses of force were also limited by laws and norms. According to Tlusty, "Men were hardly above the law just because they were on watch. Unreasonable force applied by guards was taken seriously...".[10] Within the context of the peacekeeping duties of a citizen or professional guard or militia, the nonlethal techniques of BF could provide practitioners with additional options to fulfill their duties while minimizing the risk of serious death or injury, or escalating a conflict.

[10] TLUSTY 2011: 54.

3 Practicing Nonlethal Techniques

While the precise context of BF is not known, there are many contemporaneous and plausible contexts in which the use of its nonlethal techniques would be valuable. However, it is ironic that nonlethal techniques can be more dangerous to practice today than lethal ones, given the state of modern fencing equipment—gear can protect fencers from thrusts or strikes from a Feder, but cannot protect joints from damage. For newer fencers, or those that might normally be focused on competitive longsword fencing, practicing nonlethal techniques can present a new set of challenges. In this section, we provide some advice for practitioners to safely and effectively practice locks, breaks, and throws.

3.1 Communication

Practicing nonlethal techniques can be made much safer by communicating with your partner. Communicate before starting practice. How comfortable are you with these types of techniques? Do either of you have any prior grappling or wrestling experience? Are there any injuries or limitations you should be aware of? Are there any specific techniques that they don't want to practice?

In long sword or Messer fencing, fencers ordinarily communicate *after* an exchange—whether they were hit, whether the quality was good or not, and more generally whether a technique was effective. For locks and breaks, some communication needs to occur *before* the full completion of a technique, before joint damage occurs. This could be a rapid tap on your partner or yourself, or a simple verbal acknowledgement. Conversely, one should also tell a partner whether a technique has been stopped prematurely.

After each repetition of a technique, communicate with your partner—was the technique practiced too fast or slow? Should your partner have stopped earlier or later? Constant communication after each iteration can improve your practice, and increase the safety.

3.2 Start Easy

Because of the risk of serious injury, it is best to start early repetitions of a new technique in a limited capacity. Limiting the speed, and the extent to which a lock, break, or throw is completed can let you identify any problems earlier, at lower speeds. In addition, it allows you to identify which portions of a technique can be sped up safely, and which portions should be approached with caution. Starting easy can also entail using a different simulator--a stick or an empty hand instead of a training Messer or a dagger, can reduce the number of variables that can go wrong in a technique. For throws, this could mean stopping short of a full throw, or using a crash pad. Conversely, a partner receiving a technique can encourage their partner to move slowly, and avoid providing technical feedback until it's been established that an interpretation can be practiced to completion safely.

3.3 ANATOMY IS IMPORTANT

Before attempting a lock or throw, have an idea of the basic limit to the range of motion that is being exploited. For example, if a shoulder is being hyper-rotated, get an idea of how far it will normally rotate on your partner. This will help you determine where to slow down the practice of your technique. In addition, understand that some joints are more "brittle" than others—elbows and knees are extremely "brittle," and sustain damage very quickly after they experience pain, while wrists and shoulders have more of a margin of error—and are therefore safer to use some pain as a signal for locks and submissions.

3.4 COMPLETE THE TECHNIQUE

Remember that the goal for many nonlethal techniques is usually a lock, break, or throw. Frequently, practitioners unable to successfully complete these techniques console themselves with the fact that if the technique fails, they are in a strong position to wound or kill their opponent. This reasoning implicitly ranks lethal techniques as superior to nonlethal ones. While there are some cases where a nonlethal technique is used as a setup for a lethal one, there are many plausible contexts for BF in which nonlethal techniques might be valuable, as mentioned earlier. With this in mind, strive to train safely and effectively so that nonlethal techniques are as viable an option as you as lethal ones.

4 CONCLUSION

This article provides some plausible contexts for the nonlethal techniques of Part A of BF, and some advice on practicing these techniques for fencers that primarily focus on competitive long sword or Messer fencing. The potential contexts for BF can show us why the nonlethal techniques might have been valuable to its owner and the practitioners of its techniques. Understanding these contexts can inform and shape our practice, and enrich our understanding of the martial arts of the period.

Joining the Dots: Additional Comments on the Connection Between *Bauman's Fight Book*, the *Gladiatoria* Group, and Fiore de'i Liberi

Dierk Hagedorn
Hammaborg – Historischer Schwertkampf

No two of the many surviving fighting treatises, frequently referred to by the German term *Fechtbücher* (or, anglicised, "fight books"), are exactly alike. However, there is a huge stock of individual segments, general disciplines, or personal teachings that serve as the composing material for those books. This results in a great many very special and unique specimens, but also in a vast and complicated family tree that gets more and more expansive and intertwined in the course of time as later authors or compilers made good use of older material, used it as is, rearranged it, or enhanced it with techniques of their own design—or all of the above. Thus, disentangling this intricate web can be daunting—if it's possible at all. What we, however, are definitely able to ascertain is recognising certain family characteristics and identifying specific individuals in this ever-growing family tree and putting them under our looking glass for further investigation and detailed examination. The *Gladiatoria* group of manuscripts is such a family member we shall get to know more closely.

1 The *Gladiatoria* Group

There is a considerable amount of information available that covers details about that particular group of manuscripts,[1] so I will sum it up only briefly here. The name *Gladiatoria* group designates at least three or up to nine manuscripts, depending on the presuppositions of the corresponding examination. The name itself is derived from a single manuscript which is kept in the Jagiellonian Library in Krakow (siglum GK). On its first page, the title 'Gladiatoria' (that is: belonging to swordplay) is written in green ink and Roman capital letters, thus differing from the rest of the manuscript (or any of the other manuscripts in the group) which employs a *Bastarda* script throughout. No other member of the group provides us with either the same or any other title at all. The topic of the manuscripts is mostly combat in armour, although not exclusively, painted in detailed, colourful illustrations.

---

[1] Hils 1987; Lorbeer 2006; Lorbeer et al. 2006; Leng 2008; Welle 2014; Hagedorn & Walczak 2015.

1.1 DEFINING *GLADIATORIA*

Due to the copy-and-paste nature of late medieval or Renaissance-era fight book creation, it is not possible to determine an unambiguous and crystal-clear outline of the *Gladiatoria* group since there are early and late examples that are fed from a variety of sources. What makes matters worse is that it is not even possible to determine the essential components of the *Gladiatoria* material since in itself it is not of a uniform nature; the entire *Gladiatoria* complex has distinctly ragged edges. Nevertheless, the one thing that comes to mind when the term *Gladiatoria* is mentioned is a depiction of pairs of combatants in armour—generally in a quite specific style that was the *dernier cri* in Germany in the second quarter of the 15th century.

The eponymous manuscript from Krakow, for instance, features harnesses from the transition period of so-called *Kastenbrust* harnesses with an edgy silhouette and a mostly plain surface to the heavily fluted and decorated Gothic style. This enables us to pinpoint the date of creation of this particular book in the 1440s. The armoured fighters are equipped with a number of weapons, such as spears, swords, and daggers, and they also don't shy away from grappling and wrestling. Occasionally *ecranches* are shown, small handheld shields with a characteristic indentation on the right side, the *bouche*. *Ecranches* such as these are also featured prominently in Baumann's fight book (siglum BF).

1.2 THE CORE OF *GLADIATORIA*

Three codices, all of them manufactured out of costly parchment, are quite closely related to each other; they form the core of the *Gladiatoria* group and purvey its essence. These are, next to the codex from Krakow (GK), the ones from New Haven (GN) and Vienna (GW). The overlap in contents is considerable and with only very few exceptions identical. This is true for both the images and the explanatory texts.

The armoured combat section starts with spears and *ecranches*, moves forward to swords, followed by daggers. There's also some wrestling thrown into the affair for good measure. Once we have come so far, only GK offers some unarmoured techniques with long duelling shields and swords, with sword and buckler, with *Messer* and Hungarian shield, and with staffs. The last section about ground fighting is present in all three manuscripts again.

These three core manuscripts are quite likely to stem from one single workshop, with GK being the latest exemplar. The other two, GN and GW, were created slightly earlier, judging from the style of costume and armour. They are particularly close to each other; they even share the same artist and the same scribe—at least partially.

GN has a very homogenous overall appearance; it was created by only one painter and one writer.[2] It's condition, however, is not pristine, several sheets are missing (and have been so for centuries), and occasionally the pages have been damaged to the extent that the text portions beneath the fighters' feet have been cropped. Apart from that, it can be considered as the most accomplished achievement of the entire *Gladiatoria* complex. Its endeavour to individualise each combatant is unparalleled to any earlier or later rendition of the same material. No two depicted harnesses are alike, and a vast variety of armour details can be observed.

GW on the other hand is more complete than GN content-wise, but it was produced by two scribes and four artists.[3] The first scribe is identical to that of GN, as is the third artist. The otherwise consistent appearance of GW suggests that it was created in one workshop with the contribution of several artisans.

The three core *Gladiatoria* manuscripts, GN, GW, and GK, seem to have been created within a very short timeframe of roughly ten years.

1.3 *GLADIATORIA* BY EXTENSION

There are six additional manuscripts that share material from the *Gladiatoria* complex.

The first one, manuscript BF, can be seen as a predecessor, although it most likely did not serve as a direct template.

It was a template, however, for three further volumes (PMD, PMM, and PMW), authored by a famous collector and creator of fight books, the Augsburg council servant Paulus Hector Mair (1517–1579). These books are double volumes and collect almost the entire knowledge about duel fighting that was available at the time. Mair based them on fight books he had previously acquired, the final one being the *Hans Medel* or *Sigmund Schining* manuscript (HM), bought in 1566. They are lavishly illustrated by skilled artists that belong to the circle of the famous Renaissance painter Jörg Breu the Younger (after 1510–1547). The books are excessive, both in extent as well as in expenses. In order to finance his sumptuous hobby, Mair made one grab too many in the city's treasure box, and so he was hanged as a common thief in 1579.

In 1556, he had purchased manuscript BF and incorporated large portions of it in his own books, and quite frequently they were considerably augmented by additional material or, when no text was provided, with instructions based on his own ingenuity. Consequently, the *Gladiatoria* material that is contained in BF was passed on to PMD, PMM and PMW; the three core *Gladiatoria* manuscripts GN, GW and GK did not play a role in this.

Another manuscript from Paris (P), dated to the last quarter of the 15th century, features 36 uncaptioned illustrations that also appear in the three core manuscripts.

[2] For a detailed examination of GN see HAGEDORN & WALCZAK 2015.

[3] LENG 2008: 30.

The sixth manuscript that contains a considerable amount of *Gladiatoria* material is GWo, dated to 1465–80. As another later representative, it contains some unprecedented material. About thirty images, most of them showing armoured dagger techniques, have no known counterpart. Sadly, like the Paris manuscript, none of the pictures is accompanied with text.

This raises the question whether there was a much larger corpus of *Gladiatoria* material to begin with that served as the source material to a variety of manuscripts, and the individual authors took what they liked, needed, or were requested to use. So it may be possible that the similarity of the three core manuscripts is purely coincidental.

1.4 Excursion: gauntlets and artistic convention

It is conspicuous that the armoured fighters in the three core *Gladiatoria* manuscripts don't wear gauntlets (Fig. 1, top), except for two single images in which they are addressed in the text. When we consider BF as a predecessor and a possible source, we notice gauntlets in armoured combat throughout (Fig. 1, bottom). So we may safely assume that the lack of gauntlets is a mere artistic convention which allowed the artists to portray the position of the fingers more clearly.

Manuscripts P and GWo also lack illustrated gauntlets, while PMD, PMM and PMW, feature them (just like their template BF).

This convention is by no means unique to *Gladiatoria*, it can also be observed in some of Hans Talhoffer's manuscripts.[4] As a further means of distinction, the bare hands are here mostly coloured in a light steel-blue which was also used for the rest of the harnesses. Admittedly, some occasional flesh-coloured hands do appear, possibly only due to some negligence on the side of the painter.

3 Some Connected Dots

After Rainer Welle's seminal work[5] there doesn't seem much left to say about manuscript BF. He has examined it to such an extent that he apparently has it squeezed dry. He has covered everything from a thorough codicological description to an analysis of its contents and to a graphical concordance of the images with comparable sources.[6] He also pointed out that fifteen images share a strong resemblance to manuscripts from the *Gladiatoria* group.

Upon closer inspection, I would like to add some additional ones to that list. However, a certain ambiguity remains since some of the depicted stances are of a somewhat generic nature that are rather common and likely to be assumed in a combat situation.

[4] See HTB and HTK.

[5] Welle 2014.

[6] Ibid: 132–183.

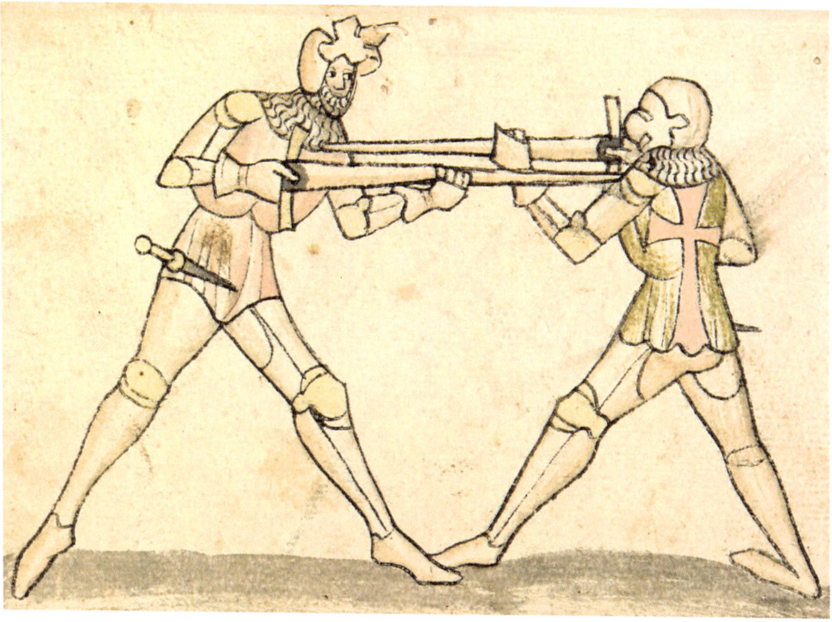

Figure 1: A piece with and without gauntlets (top: BF, f 106v; bottom: GN, f 23v).

Figure 2: Left and right binds (top: BF, f 77v; bottom: BF, f 79v).

3.1 Generic vs. Specific

When performing a martial art, there are occasionally actions that are ubiquitous, such as launching a strike with a fist, a sword, or a lance. This initial motion may be countered in various ways, and the longer the fight ensues, the more body movements are to be expected that are eventually more or less singular in appearance—depending on the physical capabilities of the fighters and, of course, their style, be it a personal one or one learned in a school.

One of those more obvious and simple generic techniques in swordplay is the moment when both combatants' blades cross and form the so-called bind. Neither of them has yet defeated their opponent, but nevertheless, both have come away unharmed—so far, that is. From this generic bind a lot of options arise, depending on pressure, distance, stillness, or motion of the antagonists. So it's a natural thing to see the bind in a manifold number of sources; in the case of BF for instance on folio 79v (right bind; Fig. 2, bottom) or folio 77v (left bind; Fig. 2, top). Other situations are remarkably unique although they don't seem to look that special on a cursory glance, such as the right fencer on folio 79v. He has adopted a position that has no equivalent to any other known source.

3.2 More Dots: Fiore de'i Liberi

It has been pointed out—mostly in online articles or blog posts that are sadly not or no longer accessible to the general audience—that there is another strong relation between the *Gladiatoria* group and yet another series of manuscripts, namely those authored by the Friualian master Fiore de'i Liberi in the first decade of the fifteenth century. Due to that overlap, three images in Fiore's armoured combat sections share the same typology with the *Gladiatoria* material presented in BF. Another three images in the unarmoured swordplay section in BF resemble Fiore's teachings also quite closely, particularly a step on the opponent's blade (f 26v) that is characteristic to Fiore and cannot be found in other sources.

The following concordance tables offer comparisons of those images from manuscript BF that have been identified as showing the same techniques in either the *Gladiatoria* group, or in Fiore de'i Liberi's books, or in both. The manuscript I chose for the *Gladiatoria* comparison was the early and artistically most exhaustive exemplar GN, and only when images were not available, I turned to GK instead. For Fiore, I relied mostly on FLG with one addition from FLP.

Some of the images are mirrored and sometimes techniques are shown from the other side, as if the portraying artist demi-circled the fighters in order to get a clearer, or just an alternative view.

Unarmoured Swordplay

BF: 76r

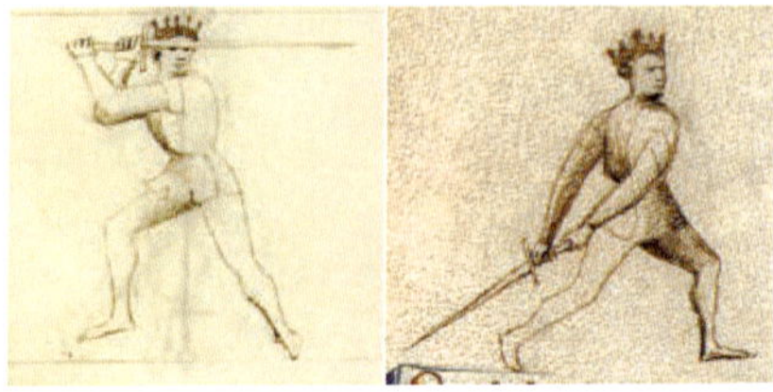

FLG: 25v-C, 26v-A

BF shows a combined stance from two of Fiore's fighters; it takes the legs from one and the upper body from another. Both combatants in BF mirror each other's stance.

BF: 76v

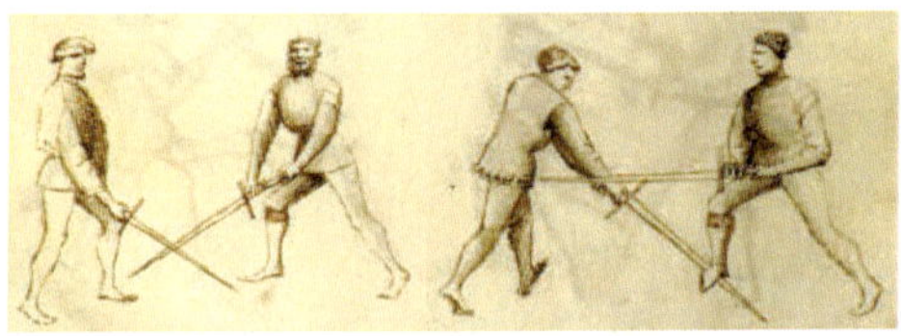

FLG: 28v-C/D

Fiore's break of the thrust. The right fighter's sword is still in a bind, but he has already put his foot on his adversary's blade.

BF: 77r

FLG: 26r-D, 26r-A

The two fighters in BF assume the same guards as those from FLG, although shown from the other side. The right fencer has also the other leg in front.

Armoured combat

BF: 81v, left

Identified by Welle.

GN: 29v

BF: 81v, right

Identified by Welle.

GN: 22v

BF: 83v

Identified by Welle.

GK: 39v

BF: 84r, right

Identified by Welle.

GN: 9r

BF: 84v

GN: 10r

Mirrored images.

BF: 86r

GK: 43v, FLG: 13r-A

Identified by Welle. Fiore shows the same technique without armour.

BF: 88r, left

GK: 47v

Identified by Welle.

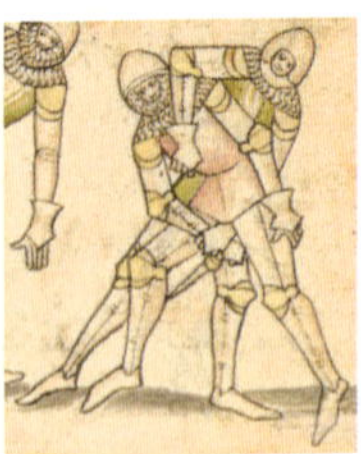

BF: 88r, right

GN: 39r

Identified by Welle.

BF: 88v

GN: 4r

Identified by Welle. The stances are similar although not exactly identical.

BF: 90v, left

GN: 28v

Identified by Welle. Mirrored images.

BF: 90v, right

GN: 27r

Identified by Welle. Mirrored images.

BF: 91r, left

FLG: 34v-B

One of Fiore's essential guards with the sword in armour.

BF: 91v **GN: 26r**

A rather similar technique, but while the left fencer in BF uses a dagger, his equivalent in GN stabs the opponent's foot with a sword.

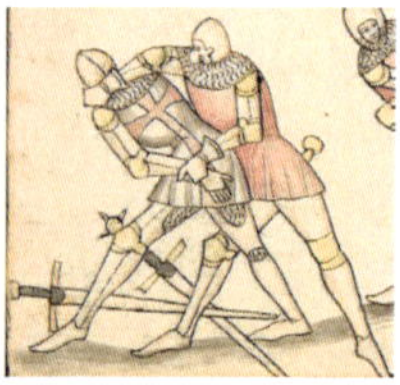

BF: 93r, left **GN: 21r**

Identified by Welle.

BF: 94r, left **GN: 29r**

In GN the text advises the fighter to grab over the opponent's elbow while the image has the hand on the head.

BF: 94r, right **GN: 21v**

Identified by Welle.

BF: 95v

GN: 6v

Identified by Welle. While the thrusts of the left fighters are similar, the targets are decidedly different ones.

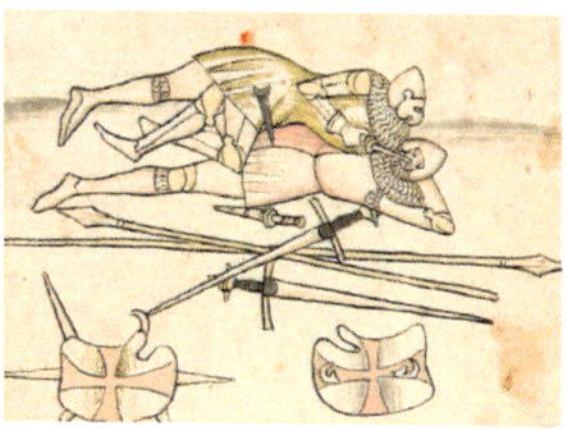

BF: 103r

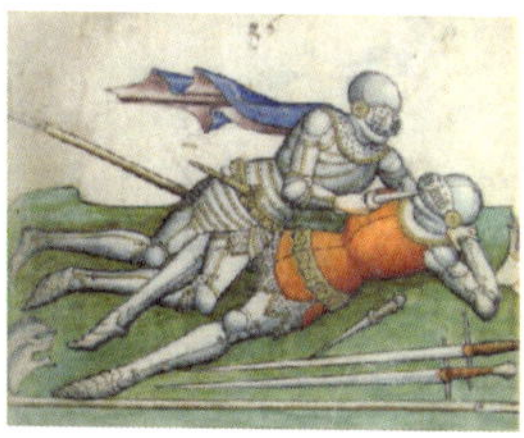

GN: 43r

Identified by Welle.

BF: 105r, left

GN: 4r, left

Identified by Welle.

BF: 105r, right

GN: 5v, left; FLG: 34v-C

Mirrored stances in BF and GN or FLG respectively.

BF: 105v

GK: 2r; FLP: 8r-A

Similar but not entirely identical mirrored stances.

BF: 106r

GN: 15r

Rather similar techniques although the target area in BF is the face while in GN it is the inside of the elbow.

BF: 106v

GN: 23v

Identified by Welle.

Long shields

BF: 96r **GK: 51v**

Similar, although not identical. GK shows the situation from the other side, and also both shields are upright.

BF: 97r **GK: 49v**

A similar situation, although BF features a judicial combat with clubs according to Franconian law, while GK shows swords following Swabian law.

BF: 97v **GK: 52v**

Mirrored stances, showing a bind with the long duelling shields.

It den gheerste steke van bove se zalmen
hem vanghen en trecke hem starc neder
het weset daer af se nemt zijn dagghe mett
armen en treckse hem om oft crijcht hem met
zijne elleboghe en draeyt hem om in zijn buuck
ze

Appendix: A Transcription and English Translation of *Vechtboek*, ms. BPL 3281

Reinier van Noort
Ense et Mente, Fekteklubben Frie Duellister

Oskar ter Mors
Virtual Fechtschule, HVN, Ludolph van Ceulen

In this article, we will present a new transcription and English translation of a manuscript fragment held in the collection of the Leiden University Library under shelf mark BPL 3281, known as *Vechtboek* (V). The manuscript consists of sixteen pages of roughly 150 × 190 mm, each containing one drawing of a dagger fighting technique, with a short descriptive text underneath each drawing. The manuscript fragment has been dated to ca. 1520–30. The pages seem to have been bound after writing, as they have no logical order.

Furthermore, we distinguish three different hands in the production of this manuscript. However, it is not possible to say with any certainty whether that means that it was produced by three different writers, or by the same person over a longer period of time.[1] While there are some striking differences in color between the illustrations, the only correlation between writing hands and illustrations, is that the illustrations with the first hand use lighter colors than the rest. This might indicate that the illustrations were made by yet another person, or that the illustrations and text were produced at different times, using different materials.[2] With regards to the origin of V, the archivist of the Leiden University Library simply asserts that linguistic characteristics point to the Western Netherlands.[3] The dialect

---

1 The first hand produced pages 1–4, 6 and 7, and can be distinguished by its short and curvy pen strokes. The second hand produced pages 5 and 8–12, showing very pronounced downward strokes and generally darker letters. The third hand, which produced pages 13–16, looks very similar to the first, but uses different spelling and abbreviations.

2 The illustrations on the pages that were written by the first hand are much lighter in colour than the others. The restoration report assumes that this is because the first illustrations are faded, but in that case, we should also expect the text and outlines to be faded, which is not really the case. Additionally, in all illustrations the artist has used shading lines, which can mostly be found in the lighter yellow used throughout. Wherever we see the lighter red, a large number of lines can be found as well, whereas the darker red only rarely sports shading. This would indicate that illustrations found on pages 1–4, 5 and 6 were intended to have this somewhat lighter colour. While all lighter illustrations are found on pages written by the first hand, there are no clear distinguishing features between the darker illustrations on pages written by either the second or the third hand.

3 The BPL 3281 item also includes a typed note by the archivist describing and superficially analyzing the manuscript.

in which this text was written, particularly through words such as *owen* (*uwen*) and *ierst* (*eerst*), seems to indicate an origin in Brabant, Holland or Flanders. Furthermore, as the text does not display some of the characteristics of the latter two dialects, we can further narrow down the origin to Brabant.[4] As the major cities in Brabant at the time the manuscript was written, such as Brussels, Bergen op Zoom, Antwerp, 's Hertogenbosch, and Malines, all had fencing guilds, it is not unlikely that the manuscript originated in any of these cities.[5]

As noted by GEVAERT & VAN NOORT, the illustrations in V are similar, artistically, to the illustrations showing dagger fighting in part A of *Bauman Fight Book* (BF), as noted by ŻABIŃSKI & WALCZAK.[6] The fighters also wear similar clothing, have similar hairstyles, and use similar daggers, though the daggers in BF have hilts in the same color as their blades (indicative of steel hilts), and were identified by ŻABIŃSKI & WALCZAK as rondel daggers with "round, slightly convex pommels",[7] while those in V have yellow hilts, suggesting they are made of brass or a similar alloy, and have wheel pommels.

Despite these artistic similarities, there does not appear to be a strong overlap in martial technique shown in the two works. Compared to the section in BF, V shows a stronger focus on the use of the forward grip. Of the sixteen attackers shown in V, eight use a forward grip, and eight use a backhand grip. These attackers are met by seven unarmed defenders (three against a forward-gripping attacker, and four against a backhand-gripping attacker), six defenders using a forward grip (five against a forward-gripping attacker and one against a backhand-gripping attacker) and three defenders using a backhand grip (all against backhand-gripping attackers).[8]

This stronger focus on the use of a forward grip, in combination with defenses against opponents grabbing your collar or hair shown in V, may indicate that V deals primarily (or exclusively) with unarmored fighting whereas BF was written with fighting in armor also kept in mind.[9] Furthermore, hair-grabbing in other fight books often indicates earnest fighting or self-defense, as such techniques

4 LOEY 1949.

5 GEVAERT & VAN NOORT 2016: 376, 380, and 381.

6 ŻABIŃSKI & WALCZAK 2002.

7 IBID: 3.

8 See GEVAERT & VAN NOORT: 392–393. Note that GEVAERT & VAN NOORT must have made an error here, as they report 18 forward gripping fighters, 7 unarmed fighters and 7 backhand-gripping fighters. By our count, unarmed defenders are seen against forward-gripping attackers in plates 1, 2, 14; and against backhand-gripping attackers in plates 5, 8, 9, 12. Forward-gripping defenders are seen against forward-gripping attackers in plates 3, 6, 10, 15, 16; and against a backhand-gripping attacker in plate 4. Backhand-gripping defenders are seen against backhand-gripping attackers in plates 7, 11, 13. This gives V a total of 14 forward-gripping fighters, 11 backhand-gripping fighters, and 7 unarmed fighters, whereas BF displays 9 forward-gripping fighters (6 of which are defenders), and 19 backhand-gripping fighters.

9 As also proposed by GEVAERT & VAN NOORT: 392–393.

Figure 1: A hair pull in wrestling (Petter, p 18).

would not be acceptable in more playful context. Prominent examples of this are found in the writings of Nicolaes Petter, Johann Georg Pascha, and the anonymous wrestling master of SR. It follows that at least some of the plays from V were likely meant for use in a self-defense context.[10]

In the following transcription, *italics* have been used to expand contractions. In the transcription, uncertainties have been placed between angle brackets ("‹...›"). These brackets have also been reproduced in the translation to indicate these uncertainties in the transcription. Square brackets ("[...]") in the translation indicate additions to the text made by the translators to improve clarity and readability. Where this was considered necessary, further comments on the transcription and translation have been added in footnotes.

10 In the second part of Petter (see Van Noort 2019), at the end of an exchange that starts with a push to the chest and ends up on the ground, he shows a ground-wrestling technique that involves hair-grabbing. In the fourth part of the same work, Petter shows an exchange that starts with one grappler grabbing the other by the hair, and ends with that other grappler being thrown over the back of first, pulled by his hair. The anonymous master from SR teaches a hair-grab that ends with running the victim head-first into a wall (see Tobler 2001: 203). German fencing master Johann George Pascha likewise includes grabbing the opponent by the hair to gain an advantage in four of his grappling lessons (see Van Noort 2018).

01r It*em* staec ieme*nt* na u van onder arme*n*
soe legt u rechte hant over u slincke ghecrust
tret in mette*n* rechte en*de* volcht mette*n* mette*n*
slincken na en*de* tret noch eens om*me* en*de* soe
worptten tegen den gront oft stect u hant diep
doere toet over zijne elleboghe en*de* stect hem zijne
hoeft tussen u beene*n* en*de* clapten op zijne ‹hoefte›
en*de* stecten in eene*n* ‹sack›.

If someone thrusts at you below [your] arm, then place your right hand crosswise over your left, step in with your right [foot] and follow with your left, and step around once more, and in this way throw them to the ground, or thrust your hand through deep, over and past their elbow, thrust their head between your legs, and strike on their ‹head› and stick them in a ‹sack›.

02r It*em* oft Ieme*nt* na u stake van onder soe legt
u handen den over dander u rechte hant bove*n*
datme*n* owen eene*n* Duim niet en zije en*de* treet
om*me* mette*n* rechte*n* bene en worpt hem den erm
op u scower en*de* nemt hem zijn Dagge oft slaetse
hem inden hals en*de* tort sterk om*me* i oft ij vers

If someone thrusts at you from below, then place your hands over each other, your right hand on top, such that you cannot see your one thumb. Step around with your right leg, and throw their arm onto your shoulder and take their dagger, or strike them in the neck and twist around strongly, one or two turns.

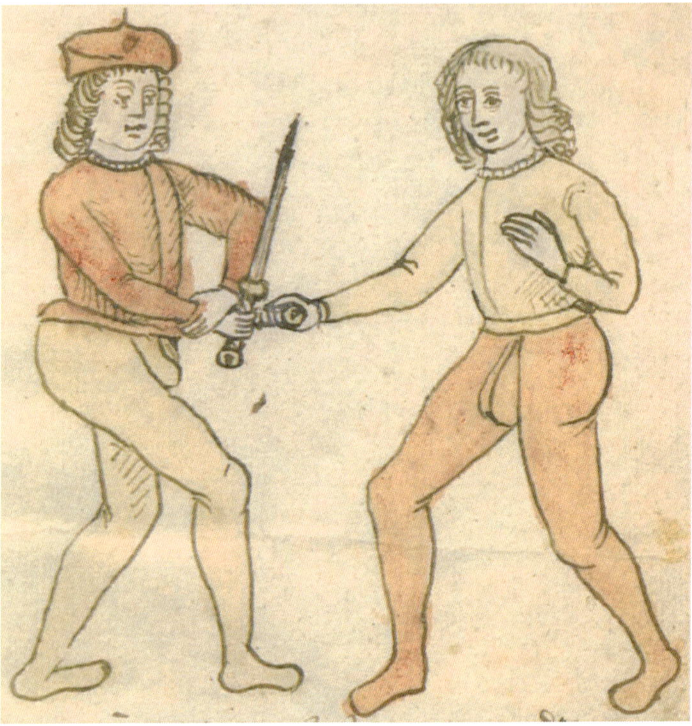

03r It*em* oft ieme*nt* na u ſtake van onder
soe worp u Dagge tegen zijne*n* ſteck en*de* werpse
verre van u oft nemtse hem met uwer
clincker hant en*de* alsoe hebdi beide Daggen.

If someone thruſts at you from below, then throw your dagger againſt their thruſt and throw it far from you, or take [their dagger] from them with your left hand and then you have both daggers.

04v It*em* den ghecruſte*n* ſteck van bove*n* soe zalmen
hem vanggen en*de* trecke*n* hem ſterck neder
het verset daer af soe nemt zijn Dagge mett*en*
⟨cruine*n*⟩ [11] en trecse hem om*me* oft grijp hem met
zijne elleboghe en*de* duwet hem om*me* in zijne*n* buick

The crosswise thruſt from above: you will catch this and then pull it down ſtrongly. The counter to that: take their dagger by the ⟨pommel⟩ and pull it around, or grab them at their elbow and push it around into their belly.

11 This could possibly be supposed to read *duime*, which would change the translation to "take their dagger with [your] thumb".

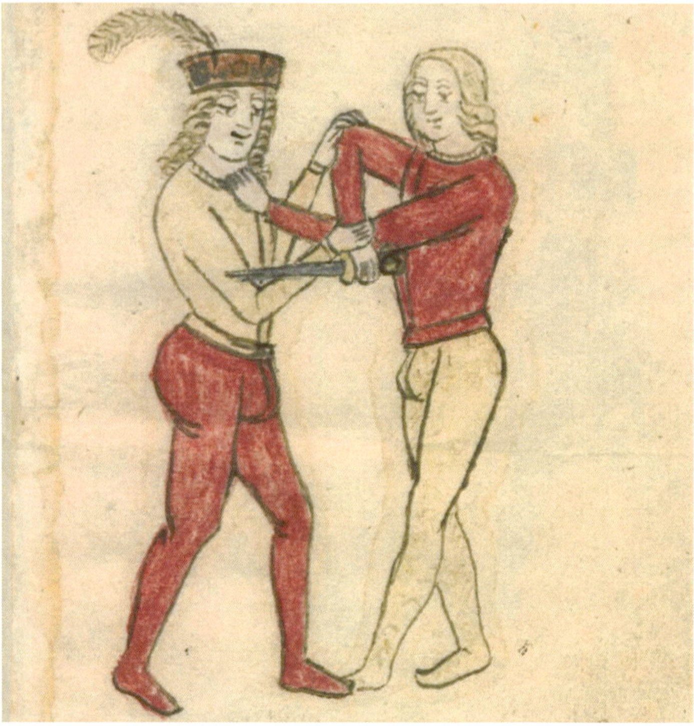

05r It*em* waert dat ieme*nt* na u stake van boven
en*de* hi u mette*n* collijere grepe soe ~~nemt~~ grijpt
hem zij vuist neeme*n* zijn Dagghe met ower
rechte hant en*de* met ower sclincker hant aen
~~owen~~ zijne*n* elleboghe doute*m* sterck van u en*de*
tort dobbel om*me* en*de* alsoe zijde hem quite en hebdi
zijn Dagghe

If someone thrusts at you from above and they grab you by your collar, then grab their fist. Take their dagger with your right hand and with your left at their elbow push strongly from you and twist around doubly. Thus, you will be rid of them, and have their dagger.

06r

It*em* oft ieme*nt* na u ſtake van ondere
soe zeldi u Dagghe in beide u handen
nemen en*de* worpen hem u Dagghe vore
en laten hem daer tegen ſteken en*de* gripen
hem dan zijn duime*n* ghe ⟨v⟩ aggen[12] gheheeten
het prisoen

If someone thruſts at you from below you will take your dagger in both your hands and throw your dagger in front of them to let them thruſt againſt it. Then capture their thumb. This is called "the Prison".

12 It is somewhat unclear whether this word is spelled as *ghevaggen* or *ghenaggen*. Regardless, this should most likely read *ghevangen*, which is translated as "captured".

07r It*em* oft iement na u ſtacke van bove*n* gheheete*n*
den haec soe zalment hem witer hant trecke*n*
en*de* leggen owen duim optte Dagghe bove*n* opte
cruine en ont haecse hem maer ghi moet
owen rechten woet om*me* ſtellen en*de* onthaecse
hem en*de* valt met ower clincker hant op
zijne*n* rechte*n* erm en*de* druſte*n* toeter eerden neder

If someone thruſts at you from above, named the hook, then you will pull it from their hand and place your thumb on top of the dagger, on the pommel and unhook it, but you muſt set your right foot around and unhook them and drop your left hand onto their right arm to push it down to the ground.

08r It*em* oft iement na u scloeghe oft na u stake
van boeven soe nemt hem zijn hant daer hi met
stect oft claet van onder zijn dagghe met ower
clincker hant en*de* met ower rechter hant stect
tussen zijn benen en*de* trect sterck neder zijn
hant daer hi die Dagghe in heeft en*de* alsoe
woerpten op u skowerre en gate*n* met striken

If someone strikes at you or thrusts at you from above then take the hand with which they thrust or strike with your left hand, below their dagger, and with your right hand thrust between their legs and strongly pull down the hand in which they have their dagger. And thus throw them onto your shoulders, and take them for a walk.[13]

13 In this sentence, *striken* has multiple possible meanings. One may simply be "to stretch", which would lead to "and stand up straight". Another potential meaning is "to go with haste", which could lead to "and move quickly". Our choice to go with "to take someone for a walk" draws an interesting parallel to a similar technique presented in Part 9 of Petter.

09r It*em* oft ieme*nt* na u stake van bove*n* soe zalme*n*
hem onttreden mette*n* rechten woete en*de* gripen
met ~~met~~ ower clincker hant aen zijne*n* rechten
eerm u iiij vinggers bove*n* en*de* owen duim onder
owen elleboghe in zijn zijde en*de* gripen hem dan
van onder zijn rechte been van onder zijn knie en*de*
en*de* licht zijn ben oppe en*de* alsoe moet hij wallen
perfortse

If someone thrusts at you from above, you will step away from them with your right foot and grab their right arm with your left hand, your four fingers above and your thumb below. [Put] your elbow in their flank and then grab them under their right leg, under their knee, and lift their leg up. And thus they must fall hard.

10r It*em* oft ieme*nt* na u stake van onder soe worpt
hem u Dagghe tegen zijnen zijne steck en*de* claecse
hem witer hant laet u een hant over de zijne
loepen en*de* nempse hem in ij mannieren af

If someone thrusts at you from below, then throw your dagger against their thrust and strike it out of their hand. Let your one hand run over theirs and take [their dagger] from them in two manners.

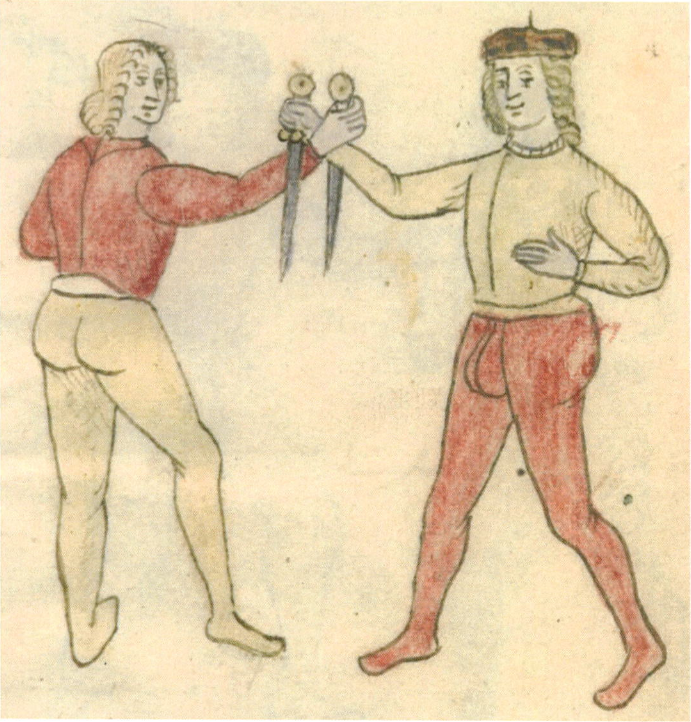

11r It*em* waert dat ieme*nt* na u ſtake van bove*n*
gheheeten den hack soe legt owen duim boven
opte cruine van ower daggen en*de* teert achterwert
om*me* en onthaecse hem tort een ⟨e⟩ oft ij vers om*me*
en*de* worpt hem zijn Dagghe witer hant.

If it happens that someone thruſts at you from above, named the hook, then place your thumb on top of the pommel of your dagger and turn around backwards and unhook it. Twiſt one or two turns around and throw their dagger out of their hand.

12r Item oft iement na u ſtake van boven ende dat hi
u grepe metter collijeren soe legt u hant opte zijne
ende sclaet van onder oppe ende grijpt hem zijn
hant daer hi zijn Dagge in heeft ende trecse ſterc
na u ende metter ander hant aen zijnen elleboghe
ende tert ſterck omme ſteſt hi van onder soe legt
u hant op owen buick ende nemtse hem ghelijc
den ghecruſten ſteck witer hant.

If someone thruſts at you from above while grabbing you by the collar, then place your hand on theirs, and ſtrike up from below. Grab the hand in which they have their dagger, pull this ſtrongly to you, with your other hand at their elbow, and twiſt [their elbow] around ſtrongly. If they thruſt from below, then place your hand on your belly and take [their dagger] out of their hand juſt as againſt the crosswise thruſt.

13r It*em* den ghecrusten steck van boven soe zalmen
hem vanggen en*de* trecken hem sterck nederre het
verset daer af soe nemt zijn Dagghe metten
hechte en*de* dous hem contrari om*me* in zijnen
buick ocht dout hem zijne*n* elleboghe contrari
om*me* en alsoe nemt hem zijn Dagge

The crosswise thrust from above: you will catch this and pull it down strongly. The counter to that, take their dagger by the grip and push it around in the opposite direction, into their belly. Or push their elbow around in the opposite direction, and thus take their dagger from them.

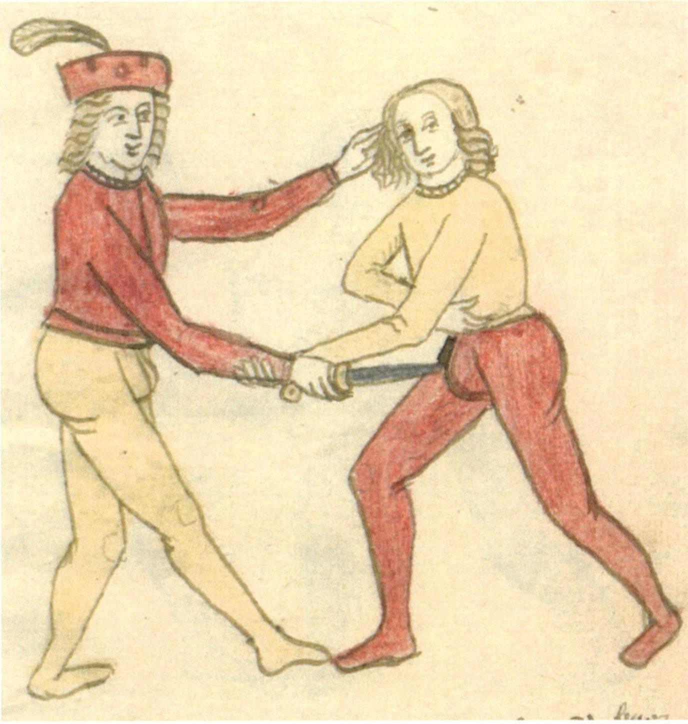

14r Ite*m* vaert bij alsoe dat u iement in u haer
grepe en*de* na u ſtak soe zalmen hem ierſt
den ſteck versetten en*de* gripen zijn Dagghe daer hi
met ſtect en*de* versclaen hem zijn hant van bove*n*
neder wert die hi in u haer heeft

If it happens that someone grabs you by your hair and thruſts at you, then you will firſt displace their thruſt and grab the dagger with which they thruſt, then beat the hand that they have in your hair down from above.

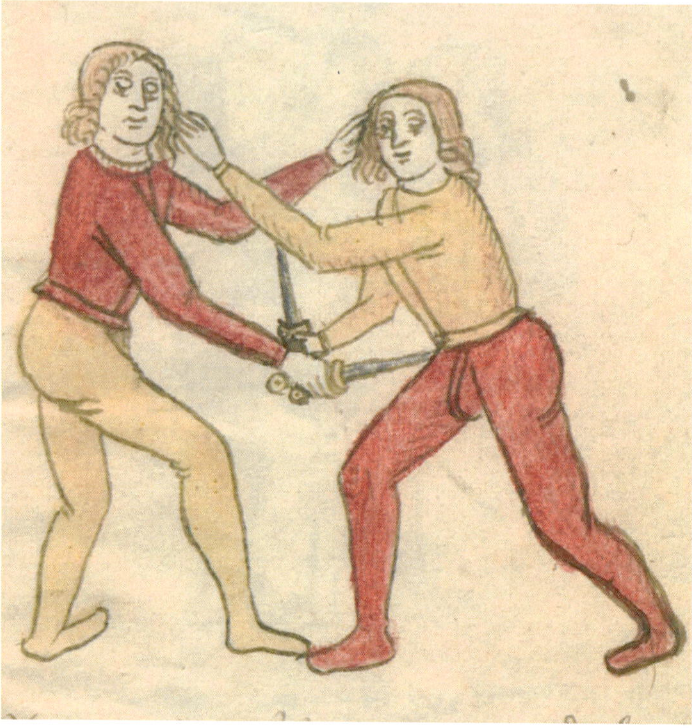

15r

It*em* waert dat elck een Dagghe inder hant
hadde en*de* grept mekanderen int haer soe wie
dat ierſt ſtake die hadde zijn Dagghe verloeren
en*de* alsoe hebdi beide Daggen

If it happens that each has a dagger in hand and grabs the other by the hair, then the one who thruſts firſt will have loſt their dagger, and you will have both daggers.

16r It*em* waert dat elck een Dagghe inder
hant hadde en*de* grepen makanderen int
haer soe wie dat jerst stake die hadde sijn
Dagghe verloeren. It*em* waert dat bij also datse
beide die naelde staken soe moetmen gade
claen en versetten hem zijne*n* steck en*de* metter
zelver hant soe moet manse uiter hant neme*n*.

If it happens that each has a dagger in hand and grabs the other by the hair, then the one who thrusts first will have lost their dagger. If it then also happens that they both thrust the needle, then you must observe and displace their thrust, and then with the same hand take [their dagger] out of their hand.

BIBLIOGRAPHY

1 PRIMARY SOURCES

1.1 MANUSCRIPTS

This table is modified from a manuscript cataloging system developed by DIERK HAGEDORN.

AD *Albrecht Dürer* or *Oplodidaskalia*
Wien, Albertina (Graphische Sammlung), Hs. 26-232

AR *Anton Rast* (AR)
Augsburg, Stadtarchiv, Reichsstadt Schätze Nr. 82

BF *Baumann's fight book*
Augsburg, Universitätsbibliothek, Cod.I.6.4°.2

BP *Berlin picture book*
Berlin, Staatsbibliothek Preußischer Kulturbesitz, Libr. pict. A. 83

FLG *Fior di Battaglia* or *Fiore de'i Liberi-Getty*
Los Angeles, J. Paul Getty Museum, Ms. Ludwig XV 13

FLP *Florius de Arte Luctandi* or *Fiore de'i Liberi-Paris*
Paris, Bibliothèque nationale de France, Mss. Latin 11269.

FLW *Die Blume des Kampfes-Wien*
Wien, Österreichische Nationalbibliothek, Cod. 5278

G *Goliath*
Kraków, Biblioteka Jagiellonska, Berol Ms. Germ. Qu. 2020

GE *Gregor Erhart*
Glasgow, R. L. Scott Collection, E.1939.65.354

GF *Glasgow fight book*
Glasgow, R. L. Scott Collection, E.1939.65.341

GK *Gladiatoria-Kraków*
Kraków, Biblioteka Jagiellonska, Berol Ms. Germ. Qu. 16

GN *Gladiatoria-New Haven*
New Haven, Yale Center for British Art, U860 .F46 1450

GW *Gladiatoria-Wien*
Wien, Kunsthistorisches Museum, KK 5013.

GWo *Gladiatoria-Wolfenbüttel*
Wolfenbüttel, Herzog August-Bibliothek, Cod Guelf. 78.2 Aug. 20.

HC *Hans Czynner*
Graz, Universitätsbibliothek, Ms. 963

HS *Hans von Speyer*
Salzburg, Universitätsbibliothek, M.I.29

HTB *Hans Talhoffer-Berlin*
Berlin, Kuperstichkabinett der Stiftung Preußischer Kulturbesitz, 78 A 15

HTG *Hans Talhoffer-Gotha*
Gotha, Forschungsbibliothek Schloss Friedenstein, Chart. A.558

HTK *Hans Talhoffer-København*
København, Det Kongelige Bibliotek, Thott 0290 2°

HTM *Hans Talhoffer-München*
München, Bayerische Staatsbibliothek, Cod. icon. 394a

JMR *Joachim Meyer-Rostock*
Rostock, Universitätsbibliothek, Mss. var. 82

JWM2 *Jörg Wilhalm-München #2* or *Lienhart Sollinger-München*
München, Bayerische Staatsbibliothek, Cgm 3712

L *Lew Fechtbuch*
Augsburg, Universitätsbibliothek, Cod.I.6.4°.3

LF *Lutegerus fight book* or *Walpurgis fight book*
Leeds, Royal Armouries Museum, FECHT 1 (formerly MS I.33)

LH *Hans Leckũchner-Heidelberg*
Heidelberg, Universitätsbibliothek, Cod. Pal. germ. 430

LM *Hans Leckũchner-München*
München, Bayerische Staatsbibliothek, Cgm 582

N *Nicolas Pol house book*
Nürnberg, Germanisches Nationalmuseum, Cod. Hs. 3227a

P *Paris Fechtbuch*
Paris, Musée National du Moyen Age, CL23842

PD *Peter von Danzig* or *Starhemberg fight book*
Rome, Cod. 44.A.8

PF *Peter Falkner*
Wien, Kunsthistorisches Museum, KK 5012

PKB *Paul Kal-Bologna*
Bologna, Universitätsbibliothek, Ms. 1825

PKM *Paul Kal-München*
München, Bayerische Staatsbibliothek, Cgm 1507

PKW *Paul Kal-Wien*
Wien, Kunsthistorisches Museum, KK 5126

PMD *Paul Hektor Mair-Dresden*
Dresden, Sächsische Landesbibliothek, Mscr. Dresd. C.93/94

PMM *Paul Hektor Mair-München*
München, Bayerische Staatsbibliothek, Cod. icon. 393

PMW *Paul Hektor Mair-Wien*
Wien, Österreichische Nationalbibliothek, Cod. 10825/6

SR *Sigmund Ainringck fight book*
Dresden, Sächsische Landesbibliothek, Mscr. Dresd. C.487

V *Vechtboek*
Leiden, Universiteitsbibliotheek Leiden, BPL 3281

1.2 Books and Incunables

Auerswald
Auerswald, Fabian von. *Ringer kunst: funf und Achtzig Stücke.* Wittemberg: Hans Lufft, 1539. <VD16 A 4051>

Monte
Monte, Pietro. *Exercitiorum Atque Artis Militaris Collectanea in Tris Libros Distincta.* Milan, 1509

Paurenfeyndt
Paurenfeyndt, Andre. *Ergrundung Ritterlicher kunst der Fechterey durch Andre paurenfeindt Freyfechter czu Vienna in Osterreich, nach klerlicher begreiffung und kurczlicher verstendnusz.* Vienna: Hieronymus Vietor, 1516. <VD16 P 1071>

Pascha
Pascha, Johann George. *Vollständiges Ring-Buch: Darinnen angewiesen wird Wie man Adversarium recht sol angreiffen sich loß machen die schläge pariren.* Leipzig: Philip Fuhrmann, 1663. <VD17 7:665262B>

Petter
Petter, Nicolaes. *Klare Onderrichtinge der Voortreffelijcke Worstel-konst, Verhandelende hoemen in alle voorvallen van Twist in Handtgemeenschap sich kan hoeden.* Amsterdam: Johannes Janssonius van Waesberge, 1674.

Rothe
Rothe, Johannes. *Ritterspiegel.* 1410

2 Secondary Sources

2.1 Books and Journal Articles

ANDREJEW & TSCHUMAKOW 2008

W. M. ANDREJEW, E. M. TSCHUMAKOW. *SAMBO – der kraftvolle, russische Kampfsport* (8th edition). Weinmann Wolfgang, 2008..

ANGLO 2001

SYDNEY ANGLO. *The Martial Arts of Renaissance Europe*. Yale University Press: 2000.

BEAUMONT 1998

NED BEAUMONT. *Kill-As-Catch-Can. Wrestling Skills for Streetfighting*. Breakout Productions, 1999.

BODEMER 2008

HEIDEMARIE BODEMER. *Das Fechtbuch. Untersuchungen zur Entwicklungsgeschichte der bildkünstlerischen Darstellung der Fechtkunst in den Fechtbüchern des mediterranen und westeuropäischen Raumes vom Mittelalter bis Ende des 18. Jahrhunderts* [Unpublished doctoral dissertation]. Universität Stuttgart, 2008.

CHAMBINAUD 2002

SEBASTIEN CHAMBINAUD. *La Lutte. Connaître – comprendre – pratiquer*. Amphora, 2002.

CHANDLER 2013

JEAN CHANDLER. "A brief examination of warfare by medieval urban militias in Central and Northern Europe". *Acta Periodica Duellatorum* **1**(1): 106–50. 2013. doi: 10.36950/apd-2013-007.

CHINAZZO 1970

LUIGI CHINAZZO. *La lotta olimpica moderna. Colpi di difesa & controcolpi, nozioni di preparazione fisico atletica*. Tipolitografia SE.PI, 1970.

COSNECK 1944

BERNARD J. COSNECK. *American Combat Judo*. Sentinel Book Publishers, Inc., 1944.

DAIGO 2016

TOSHIRO DAIGO. *Kodokan Judo Throwing Techniques*. Kodansha USA, 2016.

DÖRNHÖFFER 1907/1909

FRIEDRICH DÖRNHÖFFER. "Albrecht Dürers Fechtbuch." *Jahrbuch Der Kunsthistorischen Sammlungen Des Allerhöchsten Kaisershauses* **27**(2): 1–81. 1907/1909.

DÖRNHÖFFER 1910

———. "Albrecht Dürers Fechtbuch." *Jahrbuch Der Kunsthistorischen Sammlungen Des Allerhöchsten Kaisershauses* 27(6). Wien and Lepzig: F. Tempsky and G. Freytag, 1910.

DOROFEEVA 2019

ANNA DOROFEEVA. "Visualizing Codicologically and Textually Complex Manuscripts." *Manuscript Studies* 4(2): 334–360. 2019.

FINLEY 2021

JESSICA FINLEY. "'What's in a Name?' A Comparative Analysis of the Nomenclature of Johannes Lecküchner and Johannes Liechtenauer." *Kunst und Zettel im Messer: Bavarian State Library Cgm 582.* Somerville, MA: HEMA Bookshelf, 2021.

FORGENG 2015

JEFFREY L. FORGENG. *The Art of Swordsmanship by Hans Leckuchner.* Woodbridge: Boydell Press, 2015.

FORGENG 2018

———. *Pietro Monte's Collectanea: The Arms, Armour and Fighting Techniques of a Fifteenth-Century Soldier.* Woodbridge and Rochester: Boydell Press, 2018.

GAIN ET AL. 1980

WALTER GAIN, JÜRGEN HARTMANN, HAROLD TÜNNEMANN. *Ringen.* Berlin: Sportverlag, 1980).

GEELS 1979

ANTOON GEELS. *Judo-Bodentechnik von "Opa" Schutte.* Plus Ultra, 1979.

GEVAERT & VAN NOORT 2016.

BERT GEVAERT, REINIER VAN NOORT. "Evolution of Martial Tradition in the Low Countries: Fencing Guilds and Treatises." *Late Medieval and Early Modern Fight Books:* 376–409. Ed. DANIEL JAQUET, KAREN VERELST, TIMOTHY DAWSON. Leiden: Brill, 2016.

GRIMM & GRIMM 2004

JACOB GRIMM, WILHELM GRIMM. *Deutsches Wörterbuch, Elektronische Ausgabe, 1. Auflage.* 2004.

HAGEDORN 2017A

DIERK HAGEDORN. *Jude Lew Das Fechtbuch.* Herne: VS-Books Torsten Verhuelsdonk, 2017.

HAGEDORN 2017B

———. "Spezialeinsatz mit Mordschlag und Straße der Glieder: Der Kampf im Harnisch in den Gladiatoria-Handschriften." *'Kunst dye dich zyret' – Fechten als Mittel persönlicher und institutioneller Repräsentation.* Ed. UWE FIEDLER and THORE WILKENS. Dresden: Sandstein Verlag, 2017.

HAGEDORN 2021

———. *Albrecht Dürer - Das Fechtbuch.* Herne: VS-Books Torsten Verhuelsdonk, 2021.

HAGEDORN & TOBLER 2021

DIERK HAGEDORN, CHRISTIAN HENRY TOBLER. *The Peter von Danzig Fight Book.* Wheaton, IL: Freelance Academy Press, 2021.

HAGEDORN & WALCZAK 2015

DIERK HAGEDORN, BARTŁOMIEJ WALCZAK. *Gladiatoria – New Haven. MS U860.F46 1450.* Herne: VS-Books Torsten Verhuelsdonk, 2015.

HANNA 1986

RALPH HANNA III. "Booklets in Medieval Manuscripts: Further Considerations." *Studies in Bibliography* **39**: 100–111. 1986.

HATCHER

COLIN HATCHER. *The Flower of Battle: MS Ludwig XV13.* Ed. TRACY MELLOW. Tyrant Industries, 2017.

HENNIG 2014

BEATE HENNIG. *Kleines Mittelhochdeutsches Woerterbuch.* Walter De Gruyer GmbH, 2014.

HILS 1985

HANS-PETER HILS. *Meister Johann Liechtenauers Kunst des langen Schwertes.* Frankfurt am Main: Peter Lang, 1985.

HILS 1987

———. "'Gladiatoria'. Über drei Fechthandschriften aus der ersten Hälfte des 15. Jahrhunderts." *Codices manuscript* **1/2**. Wien: Brüder Hollinek, 1987.

HILS 1991

———. *Fecht- und Ringbuch/Vermischtes Kampfbuch. Augsburg, Universitätsbibliothek, Cod. I.6.4°2.* München: Edition Helga Lengenfelder, 1991.

HUBER & KALNING 2009

CHRISTOPH HUBER, PAMELA KALNING, JOHANNES ROTHE. *Der Ritterspiegel.* De Gruyter, 2009.

INOKUMA & SATO 1979

ISAO INOKUMA, NOBUYUKI SATO. *Best Judo.* Kodansha International, 1979.

ISHIKAWA & DRAEGER 1961

TAKAHIKO ISHIKAWA, DONN F. DRAEGER. *Judo Training Methods. A sourcebook.* Charles E. Tuttle Company, 1961.

KESLER 2016

BRANDON KESLER. *If You're Not Cheating You're Not Trying: Tips, Tricks, and Techniques for Winning in MMA, BJJ, Boxing, and Wrestling.* Self-published, 2016.

KIMURA 1974

MASAHIKO KIMURA. *Judo für Anfänger und Kämpfer.* C.A. Koch's Verlag, 1974.

KWAKKEL 2002

ERIK KWAKKEL. "Towards a Terminology for the Analysis of Composite Manuscripts." *Gazette du Livre Médiéval* **41**(autumn): 12–19. doi: 10.3406/galim.2002.1572

KWAKKEL 2012

ERIK KWAKKEL. "Late Medieval Text Collections: A Codicological Typology Based on Single-Author Manuscripts." *Reader, Author, Book: Medieval Authorship in Theory and Practice:* 56–79. Ed. Stephen Partridge and Erik Kwakkel. Toronto, Buffalo, and London: University of Toronto Press, 2012.

LENG 2008

RAINER LENG. *Katalog der deutschsprachigen illustrierten Handschriften des Mittelalters (KdiH), begonnen von Hella Frühmorgen-Voss, fortgeführt von Norbert H. Ott zusammen mit Ulrike Bodemann und Christine Stöllinger-Löser, Bd. 4/2 (Lfg. 1/2) (Fecht- und Ringbücher).* München, 2008.

LEONI & MELE

TOM LEONI, GREG MELE. *Flowers of Battle, Volume 1: Historical Overview and the Getty Manuscript.* Wheaton, IL: Freelance Academy Press, 2017.

LEXER 1885

MATTHIAS LEXER. *Mittelhoch-deutsches Taschenwörterbuch, 3. Auflage.* Stuttgart: Hirzel Verlag, 1885.

LEXER 1992

———. *Mittelhochdeutsches Taschenwoerterbuch.* Stuttgart: S. Hirzel Verlag, 1992.

LOCKE 1729

JOHN LOCKE. *Unterricht von Erziehung der Kinder.* Nicolaus Förster und Sohn, 1729.

MARSDEN

RICHARD MARSDEN. *The Flower of Battle: MS Latin 11269.* Trans. BENJAMIN WINNICK. Tyrant Industries, 2019.

MONDSCHEIN & MELE
KEN MONDSCHEIN, GREG MELE. *Flowers of Battle, Volume 3: Florius de Arte Luctandi.* Wheaton, IL: Freelance Academy Press, 2017.

MÜLLER 2011
JAN-DIRK MÜLLER. "Medieval German Literature: Literacy, Orality and Semi-Orality." *Medieval Oral Literature:* 295–334. Ed. KARL REICHL. De Gruyter, 2011. doi: 10.1515/9783110241129.295

MYSNYK, ET AL. 1994
MARK MYSNYK, BARRY DAVIS, BROOKS SIMPSON. *Winning Wrestling Moves.* Human Kinetics, 1994.

PETROV 1986
RAJKO PETROV. *Freestyle and Greco-Roman Wrestling.* FILA, 1986.

POPE 2016
BEN POPE. "Finding Safety in Feuding. Nobles' Responses to Nuremberg's Rural Security Policy in the Mid-Fifteenth Century". *VIRTUS: Journal of Nobility Studies* **23**: 11–31. 2016.

POPE 2017
BEN POPE. "Identity, Discourse and Political Strategy: Margrave Albrecht Achilles (1414–1486) and the Rhetoric of Antagonism between Town and Nobility in Upper Germany". *The Fifteenth Century XV: Writing, Records and Rhetoric:* 73–92. Ed. LINDA CLARK. Boydell & Brewer, 2017. doi: 10.7722/j.ctt1wx91t7.11

REICHL 2011
KARL REICHL. "Plotting the Map of Medieval Oral Literature." I *Medieval Oral Literature:* 3–68. Ed. KARL REICHL. De Gruyter, 2011. doi: 10.1515/9783110241129.3

ROBINSON 1980
P. R. ROBINSON. "The 'Booklet": A Self-Contained Unit in Composite Manuscripts."*Codicologica 3: Essais typologiques:* 46–69. Ed. A. GRUYS and J. P. GUMBERT. Leiden: Brill, 1980.

ROTHERT 1994
HORST ROTHER. *Ringen für Einsteiger.* Sport und Gesundheit Verlag GmbH, 1994.

ROTHERT & TEPPER 1988
HORST ROTHERT, WILLI TEPPER. *Handbuch von Grundtechniken für das Kinderringen.* FILA, 1988.

RUCH ET AL. 2014
LOTHAR RUCH, JOACHIM KÜHN, JÜRGEN SCHEIBE, JANNIS ZAMANDURIDIS. *Die Techniken im Ringen.* Meyer & Meyer Verlag, 2014.

SCHELLER & HOYLE 1995

R. W. H. P. SCHELLER, M. HOYLE. *Exemplum: Model-Book Drawings and the Practice of Artistic Transmission in the Middle Ages (ca. 900-ca. 1470).* Amsterdam University Press, 1995.

SCOTT 2019

STEVE SCOTT. *The Judo Advantage. Controlling Movement with Modern Kinesiology.* YMAA Publication Center, 2019.

SHAHMURADOV 1996

YURIY A. SHAHMURADOV. *Free Style Wrestling.* FILA, 1996.

STARBROOK 1978

DAVE STARBROOK. *Judo Starbrook Style.* Macdonald and Jane's, 1978.

THIEBAUX 1974

MARCELLE THIEBAUX. *The Stag of Love: The Chase in Medieval Literature.* Cornell University Press, 1974.

TLUSTY 2011

B. ANN TLUSY. *The Martial Ethic in Early Modern Germany. Civic Duty and the Right of Arms.* New York: Palgrave Macmillan, 2011.

TLUSTY 2016

———. "Martial Identity and the Culture of the Sword in Early Modern Germany." *Late Medieval and Early Modern Fight Books*: 547–570. Ed. DANIEL JAQUET, KAREN VERELST, TIMOTHY DAWSON. Leiden: Brill, 2016.

TOBLER 2001

CHRISTIAN HENRY TOBLER. *Secrets of German Medieval Swordsmanship: Sigmund Ringeck's Commentaries on Johannes Liechtenauer's Verse.* Chivalry Bookshelf, 2001.

TOBLER 2022

———. *Lance, Spear, Sword, and Messer: A German Medieval Martial Arts Miscellany.* Wheaton, IL: Freelance Academy Press, 2022.

VAN NOORT 2018

REINIER VAN NOORT. *The Martial Arts of Johann George Pascha.* Hagan: Self-published, 2018.

VAN NOORT 2019

———. *Nicolaes Petter's Clear instructions in the excellent Art of Grappling.* Hagan: Self-published, 2019.

WALCZAK & STARKO 2017

BARTŁOMIEJ WALCZAK, BARTOSZ STARKO. "A Survey of Dagger Teachings Among Medieval and Renaissance German Fight Books" [Poster]. *Fight Books in Comparative Perspective.* Deutsches Klingenmuseum, 9–10/11/17.

WALCZAK & STARKO 2018

———. "Additional Transmissions of Hundsfeld and Lignitzer Dagger Teachings." *Acta Periodica Duellatorum* **6**(1): 137–49. 2018. doi: 10.2478/apd-2018-0005

WASSMANNSDORFF 1870

KARL WASSMANNSDORFF. *Die Ringkunst des deutschen Mittelalters.* Leipzig: Priber, 1870.

WELLE 1993

RAINER WELLE. *«...und wisse das alle höbischeit kompt von deme Ringen.» Der Ringkampf als adelige Kunst im 15. und 16. Jahrhundert.* Pfaffenweiler: Centaurus-Verlagsgesellschaft, 1993.

WELLE 2014

———. *"...vnd mit der rechten faust ein mordstuck". Baumanns Fecht- und Ringkampfhandschrift. Edition und Kommentierung der anonymen Fecht- und Ringkampfhandschrift Cod. I.6.4° 2 der UB Augsburg aus den Beständen der ehemaligen Öttingen-Wallersteinschen Bibliothek* (2 vols). München: Herbert Utz Verlag, 2014.

WELLE 2017

———. "Ein Unvollendetes Meisterwerk Der Fecht- Und Ringkampfliteratur Des 16. Jahrhunderts Sucht Seinen Autor. Der Landshuter Holzchneider Und Maler Georg Lemberger Als Fecht- Und Ringbuchillustrator?" *Codices Manuscripti & Impressi* Supplementum 12. Purkersdorf: Verlag Brüder Hollinek, 2017.

WELLE 2021

———. *Albrecht Dürer und seine Kunst des Zweikampfes. Auf den Spuren der Handschrift 26232 in der Albertina Wien.* Graz: Sublilium Schaffer, 2021.

WENZEL 1995

HORST WENZEL. *Hören Und Sehen, Schrift Und Bild: Kultur Und Gedächtnis Im Mittelalter.* C.H. Beck Kulturwissenschaft, 1995.

WIERSCHIN 1965

MARTIN WIERSCHIN. *Meister Johann Liechtenauers Kunst des Fechtens.* München: C. H. Beck, 1965.

WINDSOR

GUY WINDSOR. *From Medieval Manuscript to Modern Practice: The Longsword Techniques of Fiore dei Liberi.* Spada Press, 2020.

ŻABIŃSKI & WALCZAK 2002

GRZEGORZ ŻABIŃSKI, BARTŁOMIEJ WALCZAK. *Codex Wallerstein: A Medieval Fighting Book from the Fifteenth Century on the Longsword, Falchion, Dagger, and Wrestling.* Boulder, CO: Paladin Press, 2002.

ZMORA 2015
HILLAY ZMORA. *The Feud in Early Modern Germany.* Cambridge University Press, 2015.

2.2 WEBSITES

HULL 2007
JEFFREY HULL. "Getting Punchy. Fist-Fighting, Wrestling and Fight-Books." *Association for Renaissance Martial Arts.* 2007. Accessed 15th August 2022. http://www.thearma.org/essays/Getting-Punchy.pdf

KIRCHHOFF 2020
KIRCHHOFF, CHASSICA. "Silver-Pommeled Swords and Old Italian Cannonballs." *Martial Culture in Medieval Town.* 2020. Accessed 15th August 2022. http://martcult.hypotheses.org/954.

LORBEER 2006
CARSTEN LORBEER. "Transkription des Tunierhandbuchs KK5013,Anonymer Verfasser, Kunsthistorischen Museum in Wien." *Gesellschaft für pragmatische Schriftlichkeit.* 2006. Accessed 15th August 2022. http://www.pragmatische-schriftlichkeit.de/KK5013.html

LORBEER ET AL. 2006
CARSTEN LORBEER, JULIA LORBEER, ALEX GIEẞL, BARBARA KAPPELMAYR, ROBERT BRUNNER, MARION FREUNDL, MARITA WIEDNER, ANDREAS MEIER, KRISTIAN BABIC. "Transkription der Fechthandschrift Ms. germ. quart. 16, "Gladiatoria", Anonymer Verfasser, Hs. Krakau, Bibl. Jagellonica." *Gesellschaft für pragmatische Schriftlichkeit.* 2006. Accessed 15th August 2022. http://www.pragmatische-schriftlichkeit.de/msgermquart16.html

VAN LOEY 1949
A. VAN LOWY. "Middelnederlandse spraakkunst. Deel II. Klankleer." *The Digital Library for Dutch Literature.* 1949 Accessed 15th August 2022. http://www.dbnl.org/tekst/loey001midd03_01/loey001midd03_01_0010.php

"Augsburg Group." *Wiktenauer.* Ed. MICHAEL CHIDESTER. Accessed 15th August 2022. http://wiktenauer.com/wiki/Augsburg_Group

"Bauman Fechtbuch (Cod.I.6.4°.2)." *Wiktenauer.* Ed. MICHAEL CHIDESTER. Accessed 15th August 2022. http://wiktenauer.com/wiki/Bauman_Fechtbuch_(Cod.I.6.4°.2)

"Johan Liechtnawers Fechtbuch geschriebenn (MS Dresd.C.487)." *Wiktenauer.* Ed. MICHAEL CHIDESTER. Accessed 15th August 2022. http://wiktenauer.com/wiki/Johan_Liechtnawers_Fechtbuch_geschriebenn_(MS_Dresd.C.487)

About the Contributors

Michael Chidester is the Editor-in-Chief of Wiktenauer and owner of HEMA Bookshelf. He is a Research Scholar of the Meyer Freifechter Guild, a founding member of the Society for Historical European Martial Arts Studies (SHEMAS), and a Lifetime Member of the HEMA Alliance. He has lectured on historical martial arts across North America and Europe and authored several books, including *The Illustrated Meyer*, *The Long Sword Gloss of GNM Manuscript 3227a*, and *The Flower of Battle: MS M 383*. Michael has been studying HEMA since 2001 and currently trains at Athena School of Arms.

Jessica Finley has had a sword in her hands for the past twenty years. Her initial interest was in stage combat, but not too long after beginning that pursuit began to ask "...but how did they really fight?", and from that question branched out to German Medieval Martial Arts. She is the head instructor at Ritterkunst Turhalle in Lawrence, Kansas. She has taught and competed internationally at events including the *Western Martial Arts Workshop*, *Paddy Crean International Workshop*, *Swordfish*, and *Longpoint*, and has taught intensives at various events and schools as well as weekend private intensives at her home. She also has a background studying Judo under the tutelage of Arden Cowherd of Topeka Judo Club. Her previous titles include *Medieval Wrestling: Modern Practice of a Fifteenth Century Art*, which is an interpretation of Master Ott's wrestling treatise of German wrestling techniques. She also researches medieval clothing construction and fabric armor and has presented her findings at the *International Congress on Medieval Studies* at Kalamazoo and published an article in *Medieval Clothing and Textiles* (Boydell and Brewer) on her study of a 15th century quilted armor. Jessica currently lives in Lawrence, KS with her family. In between "kid-wrangling" her children and remodeling her house, she writes books and does her assigned physical therapy exercises.

Dr. Elias Flatscher, PhD (born in South Tyrol/Italy, 1988) has trained and competed in Greco-Roman wrestling (mostly modern Nuremberg school) and Freestyle wrestling (mostly Dagestani school) in Germany and in the Austrian Bundesliga (seasons 2013–17), as well as training and occasionally competing in Judo, SAMBO, Sumo, Ranggeln and Schwingen. He thinks of himself as a wrestler who reads treatises—rather than a HEMAist who loves wrestling. He has coached grappling for a variety of martial disciplines (youth wrestling, MMA, BJJ, HEMA) and conducted several international workshops on medieval wrestling, so far focusing on the works of Fiore, Monte, and the "Blume des Kampfes" and the "Nuremberg" groups of treatises. When he's not training or wrestling (or thinking/reading/writing about it), he currently works as a post-doc in medieval archaeology at the University of Zurich. He can be reached at eliasflatscher@yahoo.it or on Facebook.

DIERK HAGEDORN works for the IT subsidiary of a large German airline for numerous creative issues such as user experience, interface, and corporate design. He is an instructor for the long sword at Hammaborg–Historischer Schwertkampf, a club for historical fencing in Hamburg; he is also a passionate wearer of armor. He has transcribed 30 of the approximately 80 surviving fight books in the German language and made them available on the website of his club or published them in book form. Among his numerous publications are editions of the manuscripts of Peter von Danzig, Lew the Jew, Hans Talhoffer, and Jörg Wilhalm.

MARLENE HURST began her HEMA practice in 2016 with Capital Kunst des Fechtens (CKdF) where she is now an instructor for Liechtenauer Longsword and Von Auerswald *Ringen*. She has guest lectured at DanceFight 2022, Fecht Yeah 2019, and participated in Esfinges' 2020 Instructor Symposium. She is a decorated practitioner, and is one of only a handful of women to referee *Ringen*, having done so at Longpoint in 2019.

LUCIEN LEE is the founder and principal instructor of Bastion HEMA in Singapore, one of the largest HEMA schools in Asia. He studied and taught at the Glamorgan School of Arms in Wales under the guidance of DIRK NEUMANN. In addition to more than a decade of HEMA experience, LUCIEN has worked as a Mandarin to English translator, focusing on legal contracts and novel localization. In his free time, LUCIEN enjoys watching "Sarah and Duck" with his daughter.

CHARLES LIN has been studying the *Kunst des Fechtens* of Johannes Liechtenauer since 2015. He uses historical and archaeological research, as well as a broad background in martial arts and dance, to inform his studies into the context and practice of *Kunst des Fechtens*. He is a financial economist and works in Washington, DC.

OSKAR TER MORS is a medieval historian from Leiden, the Netherlands, specialized in dueling and urban culture. Currently working as a history teacher, he still occasionally finds some time off from work and HEMA-practice to do fencing related research. Although OSKAR practices a large number of weapons and does so by working from a multitude of fight books, most of his work tends to focus on Lecküchner's *Kunst und Zedel ym Messer*, such as the full interpretation of the manuscript that was published on YouTube. Additionally, OSKAR has worked on several archaeological experiments as an expert user.

REINIER VAN NOORT started training Historical European Martial Arts in 2005, and has been researching and teaching rapier fencing since 2010. While his main focus is on the single rapier as used in the Netherlands and Germany in the late 17th century, he is also interested in any other sources in Dutch, or with a connection to the Netherlands. Currently, REINIER teaches and fences with the Fekteklubben Frie Duellister in Oslo (Norway), and shares his research and translation work mostly through Ense et Mente. As part of this research, Reinier has produced over 20 translations of fencing texts, has published more than 10 books, and has co-authored several peer-reviewed articles or chapters. Reinier has been awarded the Best Researcher Award at the 2014 and at the 2016 HEMA Scholar Awards.

Christian Henry Tobler is Principal Instructor for the Selohaar Fechtschule, a school for historical fencing, and is the author of many books on German medieval martial arts, including *Fighting with the German Longsword* and *In Service of the Duke*. His most recent title is *Lance, Spear, Sword, and Messer: A German Medieval Martial Arts Miscellany*. He has taught at numerous events in the US and abroad and is a founder of the Chivalric Martial Arts Association. He lives in Connecticut with his wife, along with a considerable collection of reproduction arms and armour. He has worked for many years, in various capacities, in the analytical instrumentation industry.

Bartłomiej Walczak has been practicing and researching HEMA since 1999. His main interests include methodology of reconstruction, nature of Medieval combat, and unravelling mysteries hidden in the manuscripts. Bart has written or coauthored several books and articles on these topics, including coauthoring *Codex Wallerstein: A Medieval Fighting Book from the Fifteenth Century on the Longsword, Falchion, Dagger, and Wrestling* with Grzegorz Żabiński in 2002.

www.ingramcontent.com/pod-product-compliance
Lightning Source LLC
LaVergne TN
LVRC081519210926
848486LV00020B/411